The Idea of the World

A multi-disciplinary argument for
the mental nature of reality

The Idea
of the World

A multi-disciplinary argument for the mental nature of reality

Bernardo Kastrup

Foreword by Menas C. Kafatos,
Fletcher Jones Endowed Chair Professor of Computational
Physics, Chapman University

Afterword by Edward F. Kelly,
Professor, Department of Psychiatry and Neurobehavioral
Sciences, University of Virginia

BOOKS

Winchester, UK
Washington, USA

First published by iff Books, 2019
iff Books is an imprint of John Hunt Publishing Ltd., No. 3 East Street, Alresford,
Hampshire SO24 9EE, UK
office1@jhpbooks.net
www.johnhuntpublishing.com
www.iff-books.com

For distributor details and how to order please visit the 'Ordering' section on our website.

Text and figures copyright © 2016-2018: Bernardo Kastrup
Foreword copyright © 2017: Menas C. Kafatos. Published with permission.
Afterword copyright © 2017: Edward F. Kelly. Published with permission.

ISBN: 978 1 78535 739 8
978 1 78535 740 4 (ebook)
Library of Congress Control Number: 2017960354

All rights reserved. Except for brief quotations in critical articles or reviews, no part of this
book may be reproduced in any manner without prior written permission from the publishers.

The rights of Bernardo Kastrup as author have been asserted in accordance with the Copyright,
Designs and Patents Act 1988.

A CIP catalogue record for this book is available from the British Library.

Design: Stuart Davies

Printed and bound by CPI Group (UK) Ltd, Croydon, CR0 4YY, UK

US: Printed and bound by Edwards Brothers Malloy 15200 NBN Way #B, Blue Ridge Summit,
PA 17214, USA

We operate a distinctive and ethical publishing philosophy in
all areas of our business, from our global network of authors to
production and worldwide distribution.

Contents

Other books by Bernardo Kastrup

Rationalist Spirituality: An exploration of the meaning of life and existence informed by logic and science.

Dreamed up Reality: Diving into mind to uncover the astonishing hidden tale of nature.

Meaning in Absurdity: What bizarre phenomena can tell us about the nature of reality.

Why Materialism Is Baloney: How true skeptics know there is no death and fathom answers to life, the universe, and everything.

Brief Peeks Beyond: Critical essays on metaphysics, neuroscience, free will, skepticism and culture.

More Than Allegory: On religious myth, truth and belief.

I must ask the reader to forgive me for having ventured to say in these few pages so much that is new and perhaps hard to understand. I expose myself to his critical judgment because I feel it is the duty of one who goes his own way to inform society of what he finds on his voyage of discovery. ... Not the criticism of individual contemporaries will decide the truth or falsity of his discoveries, but future generations. Carl Gustav Jung: *Two Essays in Analytical Psychology*.

Acknowledgments

Although this book is credited to a single author, tangible and intangible contributions from many others permeate it throughout. I would like to thank, first of all, the anonymous reviewers of the academic papers that constitute the core of this work. Peer-review is often a flawed process and I have not been spared many of its shortcomings. But I happily acknowledge that most of the papers collected in this volume have been significantly improved thanks to critical feedback and suggestions from anonymous reviewers.

I am also grateful to the editorial teams of the open-access journals that originally published my papers. Fully open-access publications—commercial and university journals alike—are often maligned for allegedly low-quality standards and poor editorial processes. While this may be the case for a number of unscrupulous journals, I am glad to bear witness to the fact that there are reliable, high-quality open-access options out there. Indeed, I believe strongly that the results of academic research and scholarship should *not* be hidden behind pay-walls. As such, I feel encouraged by the realization that a strong open-access movement is a growing reality in academic publishing. It is incumbent upon researchers and scholars alike to support this movement in every way we can.

The gracious support of David Chalmers has been instrumental in the months leading up to the completion of this work. Not only has David critically reviewed key parts of my material, he has also given me the opportunity to participate—with funding from the Global Institute for Advanced Studies of New York University, which I gratefully acknowledge—in a specialized workshop late in the spring of 2017. My participation in that event, although taking place at a time when the papers collected in this volume were already either published or completed in draft form, has helped me bring the various ideas together more effectively, so to assemble a more compelling overarching argument.

Discussions I had with other colleagues have also been valuable. With the risk of leaving important names out, I would like to explicitly thank philosophers Itay Shani, Galen Strawson, Daniel Stoljar, Miri Albahari, Michael Pelczar, Barry Dainton, and Philip Goff. Neuroscientist Anil Seth has also been of much help by pointing out to me the relatively recent literature on so-called "no-report paradigms" in consciousness research.

Author and science journalist John Horgan, a valued pal, has graciously helped my ideas get visibility in the mainstream science media, despite his not necessarily agreeing with everything I have to say. This speaks volumes to John's integrity and intellectual honesty. I am also thankful to Michael Lemonick, Chief Opinion Editor at *Scientific American*, for his continuing trust.

The feedback and encouragement I received from researchers of the Division of Perceptual Studies (DOPS), University of Virginia School of Medicine, during my visit there in the spring of 2016—for which I also gratefully acknowledge funding—have been key to the effort that eventually resulted in this book. Indeed, in hindsight, I realize that it all began there, in Charlottesville. I am particularly grateful to Edward F. Kelly, not only for the visionary Afterword he wrote for this book, but also for his continuing encouragement and guidance. I am grateful too to other scholars informally associated with DOPS, for the interesting discussions we've had.

Some people have been cherished intellectual partners— fellow travelers of the same thoughtscapes I feel compelled to explore—since before the idea of this book germinated in my mind. They are comrades and friends: Rob van der Werf, Paul Stuyvenberg, Richard Stuart, Deepak Chopra, Rupert Spira, Neil Theise, Subhash Kak, Rudolf Tanzi, Jeffrey J. Kripal and many others. I am especially grateful to Menas C. Kafatos, not only for the Foreword he contributed to this volume, but also his invaluable endorsement of the ideas expressed in it.

Last but not least, those close to me know that, despite appearances to the contrary, my inner life is not an easy ride. The loving and reassuring presence of my partner, Claudia Damian, has been the solid foundation upon which I have managed to remain centered and productive, notwithstanding my inner struggles and vulnerabilities.

Foreword by Menas C. Kafatos

The Idea of the World is an unusual and challenging book in many ways. Bernardo Kastrup pulls together previously peer-reviewed articles into a coherent new manuscript. This is not usual. However, due to the rigor pursued and the fact that the articles form a continuous and evolving set of ideas, it does make a lot of sense. It is hard to find a book that would not just repeat works already published but instead build on them and present a coherent whole. *The Idea of the World* does this marvelously.

As the author indicates, his previous approach in earlier books was to provide the readers with a "felt sense of the world" he was describing. But the present book goes in a different and, at this point, correct direction, building on his previous manuscripts to provide a philosophically sound and rigorous formulation of the idealism he espouses. Starting from many empirical facts, such as the correlations of brain activity with subjective experience (which current neuroscience, overextending itself in my view, assumes to hold a causal connection); the obvious fact that we all appear to share the same world; the fact that science is even possible and, specifically, that the laws of physics operate independently of subjective or personal wishes (but do they really?); and countless other examples that seem to point to an external, physical reality, Kastrup develops a clearly proposed ontology *based on parsimony, logical consistency and empirical adequacy*, to show that appearances are just that. Metaphors only serve a secondary role to emphasize points made.

The analytic approach may appeal especially to audiences that love precision but, I believe, does not restrict the wider appeal of the book. Coming myself from a background of scientific rigor, I believe that, in the end, it is precision and consistency that matter, even if it requires some 'hard work' on the part of the reader to follow the exquisite arguments that Bernardo

1

makes and not give up easily. As Bernardo notes, "the ontology formulated here is not an expansion, but in fact a subset of the ideas" he has tried to present in his earlier works. And quite right, the rigor is needed to counter the—seemingly—rigorous metaphysics of physicalism (I emphasize 'seemingly' because I also agree with Bernardo that physicalism just does not work, for many of the reasons that the book points out).

By articulating his ontology precisely, Bernardo builds a formidable castle of potent ideas to defend idealism and to show that it is not vague, empty or "new age stuff" as many detractors and cynics, often ignorant of philosophy, claim.

The book brings together ten different articles, each of them delightful to read and serious and deep. They were published, as stated above, in peer-reviewed academic journals. The editors of the journals did not necessarily know that the articles would eventually form a whole and improved manuscript, but indeed that is what they do, fitting and assisting each other in the development of the idealist ontology.

Besides the important aspects of rigor, consistency and of clearly demonstrating that idealism is in fact more precise and powerful than physicalism, there is another important goal that I would like to emphasize here that may be lost in the shuffle of rigorous arguments: the need to provide an ontology that *holds promise for humanity and the future*, as current physicalism has clearly failed to provide meaning to life and to bridge the gap of separation that humans experience. This is despite the success of science, which, I would claim, as Bernardo clearly claims as well, has ultimately led to observations—in the form of quantum mechanics—that do not fit the (classical) physicalist worldview.

The reader is encouraged to read the Overview to understand how the arguments are presented and build upon each other, before digging into specific chapters.

In view of the successful articulation of a modern form of the ancient ontology of idealism and the hope that it carries, and because this new book by Bernardo Kastrup is not just for philosophers, quantum physicists and mathematicians-logicians, but also for the learned generalist who is looking for meaning in a world beset by strife and division, I pronounce, Bravo Bernardo, a job well done!

Menas C. Kafatos, PhD is the Fletcher Jones Endowed Chair Professor of Computational Physics at Chapman University, Orange, California, USA.

Note to readers of my previous books

Prior to the present volume, I have written six books elaborating on my views regarding the underlying nature of reality. Particularly in *Why Materialism Is Baloney* and *More Than Allegory*, in addition to a conceptual exposition I have also made liberal use of metaphors to help readers develop direct intuition for the ideas expressed. My intent was not to win a technical argument in a court of philosophical arbitration, but to evoke in my readers a felt sense of the world I was describing. As such, my work has had a character more akin to continental than analytic philosophy.

I have no regrets about it. Yet, I have also come to recognize the inevitable shortcomings of the approach. Some readers have misinterpreted and others over-interpreted my metaphors, extrapolating their applicability beyond their intended scope. Yet others have simply become overwhelmed or confused by the many metaphorical images, losing the thread of my argument. Perhaps most importantly—given my goal of providing a robust alternative to the mainstream physicalist metaphysics (Kastrup 2015: 142-146)—some professional philosophers and scientists felt they needed to see a more conceptually clear and rigorous formulation of my philosophical system before they could consider it.

The present work attempts to address all this. Starting from canonical empirical facts—such as the correlations between subjective experience and brain activity, the fact that we all seem to share the same world, the fact that the known laws of physics operate independently of our personal volition, etc.—it develops an unambiguous ontology based on parsimony, logical consistency and empirical adequacy. It re-articulates my views in a more rigorous and precise manner. It uses metaphors only as secondary aides to direct exposition. I have strived to make every step of my argument explicit and sufficiently substantiated.

This volume thus represents a trade-off: on the one hand, its mostly analytic style prevents it from reaching the depth and nuances that metaphors can convey. Parts II and III of my earlier book *More Than Allegory*, for instance, use metaphors to hint at philosophical ideas that can hardly be tackled or communicated in an analytic style. As such, the ontology formulated here is not an expansion, but in fact a subset of the ideas I have tried to convey in earlier works. On the other hand, the present volume articulates this subset more thoroughly and clearly than before, which is necessary if it is to offer—as intended—a credible alternative to mainstream physicalism.

Incomplete as the subset of ideas presented here may be, I shall argue that it is still more complete than the current mainstream metaphysics. This subset alone—as I elaborate upon in the pages that follow—should be able to explain more of reality, in a more cogent way, than physicalism. By articulating the corresponding ontology precisely, my intent is to deny cynics and militants alike an excuse to portray it as vague and, therefore, dismissible. If the price to achieve this is to write a book as if one were arguing a case in a court of law, then this book represents my case. You be the judge.

Preface

The main body of this work brings together ten different articles I published in peer-reviewed academic journals. Unbeknownst to the journals' editors, the articles were conceived, from the beginning, to eventually be collected in the volume you now have in front of you. Despite being self-contained, each was designed to fit into a broader jigsaw puzzle that, once assembled, should reveal a compelling, holistic picture of the nature of reality. This book presents the completed jigsaw puzzle. The resulting picture depicts an ontology that squarely contradicts our culture's mainstream physicalist metaphysics.

Indeed, according to the ontology described and defended here, reality is fundamentally experiential. A universal phenomenal consciousness is the sole ontological primitive, whose patterns of excitation constitute existence. We are dissociated mental complexes of this universal consciousness, surrounded like islands by the ocean of its mentation. The inanimate universe we see around us is the extrinsic appearance of a possibly instinctual but certainly elaborate universal *thought*, much like a living brain is the extrinsic appearance of a person's conscious inner life. Other living creatures are the extrinsic appearances of other dissociated complexes. If all this sounds implausible to you now, you have yet more reason to peruse the argument carefully laid out in the pages that follow.

Each of the ten original academic articles constitutes a chapter in this volume, organized so as to present an overarching argument step by step. I have added five extra preamble chapters, as well as an overview and extensive closing commentary, to weave the original articles together in a coherent storyline.

The choice to break up my argument into ten self-contained, independently published articles had three motivations. Firstly, I have been criticized for not submitting my earlier work to the scrutiny of peer-review. I take this criticism only partly to heart: peer-review can be a prejudiced process that stifles valid non-mainstream views whilst overlooking significant faults in mainstream arguments (Smith 2006, McCook 2006, Baldwin 2014). As an author whose ideas systematically defy the mainstream, I had doubts about whether my articles would receive an impartial hearing. And indeed, often they didn't. Nonetheless, peer-review can also be constructive, insofar as it provides penetrating criticisms that help sharpen one's arguments. This was my hope and, as it turns out, several of my original manuscripts were significantly improved thanks to insightful comments from reviewers. In the end, peer-review has proven to be fruitful.

Secondly, specialized articles can reach more and different people in academia than a more generic book. The articles collected in this volume span fields as diverse as philosophy, neuroscience, psychology, psychiatry and physics, each with its own academic community. By publishing the articles in journals specifically targeted at their respective communities, I hope to have reached people who will probably never hear of—or be interested in—this book as a whole.

Thirdly, by having each part of my broader argument receive the specialist endorsement that peer approval represents, I hope to deny cynics and militants an excuse to portray the ontology presented here—antagonistic to current mainstream views as it is—as dismissible.

In the interest of achieving the three goals stated above, the articles collected in this volume were originally published in journals that, at the time of manuscript submission, met the following criteria:

1. Peer-review process;
2. Open-access policy (so to safeguard my ability to make the articles available to a wider, non-academic readership);

3. Their publishers were not included in Jeffrey Beall's list of potentially disreputable open-access publishers[1] (Beall n.d.), as of its version of 12 January 2017;[2]

4. No transfer of copyright required from authors (so to safeguard my ability to republish the articles in this volume).

To the extent possible within these constraints, I have also sought broader geographical exposure for my work by publishing in journals spanning North America, Western, Central and Eastern Europe.

In order to preserve the integrity of the original peer-review process, I am reproducing the ten original articles here without any change of substance. I have only corrected the occasional typo and language inaccuracy, harmonized the terminology and ensured consistency—citation style, section and figure numbering, etc.—across the entire book. I have also consolidated all references in the bibliography at the end of this volume, so to reduce redundancy. Everything else is as it was originally published in the respective journals. Whenever I felt that an

---

1 A study published in *Science* (Bohannon 2013) concluded, "Beall is good at spotting publishers with poor quality control," although "almost one in five [of the journals] on his list did the right thing." So Beall erred on the side of being overly critical of the journals he evaluated. By contrast, the same study showed that the Directory of Open Access Journals (DOAJ), which seeks to list only credible publications, included many journals with poor quality control. Although I understand that the DOAJ has made several improvements to its processes since then, I have nonetheless elected to use Beall's 'black list' instead of the DOAJ's 'white list.'

2 This was the latest version of Beall's list available as of the time of this writing. Jeffrey Beall had then just stopped maintaining the list, so this is possibly the last version as well.

update of—or comment on—specific passages was called for, I have done so in the form of added footnotes, so to preserve the original text.

For this reason, and since the original articles had to be self-contained, some repetition of content occurs across chapters. Some readers may consider this annoying, but I think it has a positive side effect: it provides a regular recapitulation of key ideas and context throughout the book, helping the reader keep track of the overarching argument line.

Finally, because the main substance of this work can already be found in ten freely accessible articles, it is important to highlight that the value-add of this book consists in my effort to weave the articles together in a coherent storyline, building up to an overarching ontology. By downloading the original articles one can get the pieces of the jigsaw puzzle, but by reading this book one gets the overall picture the pieces form when properly connected together.

It is my sincere hope that this picture helps you come to new insights about the nature of reality.

Overview

O poesy! For thee I grasp my pen
That am not yet a glorious denizen
Of thy wide heaven; yet, to my ardent prayer,
Yield from thy sanctuary some clear air.
John Keats: *Sleep and Poetry.*

This book is divided into five main parts: Part I makes explicit the main artifacts of thought—unexamined assumptions, fallacious logical bridges, etc.—that plague the contemporary philosophical outlook regarding the nature of reality. By pointing out these seldom-discussed artifacts, I hope to establish the need for a different approach to ontology, which Part II then attempts to fulfill by formulating an idealist hypothesis. According to this hypothesis, self and world are manifestations of spatially unbound universal consciousness or mind (throughout this book, I use the words 'consciousness' and 'mind' interchangeably, in the sense of phenomenal consciousness). Part III then reviews and refutes the objections most commonly leveled against idealism. With a view to empirically substantiating the central idea of the book, Part IV explores neuroscientific evidence that corroborates the idealist hypothesis discussed in Part II. Finally, Part V discusses the psychological motivations behind our culture's adoption of the physicalist metaphysics and the implications of idealism regarding our personal relationship with the world and the meaning of our life.

Part I lays the ground for what follows by rendering explicit the thought traps that characterize academic philosophy's most popular ontologies. Chapter 2 shows how a generalized tendency to try to replace concrete reality with mere thought abstractions lies at the root of much of our confusion today. Chapter 3 then explores the more specific consequences of this tendency: it shows that predicaments such as the 'hard problem of consciousness' and the 'subject combination problem'—which have mobilized considerable intellectual effort for the past couple of decades—are illusions arising from the fallacious logico-conceptual structures of physicalism and bottom-up panpsychism, respectively.

Part II contains the book's core message: it elaborates on idealism both from a classical perspective—discussed in Chapter 5—and a quantum mechanical one—discussed in Chapter 6.

Chapter 5 starts from the assumption of a classical world of tables and chairs, wherein objects supposedly have whatever physical properties they have regardless of being observed. It then shows how a consciousness-only idealist ontology can account elegantly for this classical world. The argument in Chapter 5 reconciles idealism with our everyday intuitions about reality.

However, quantum theory and the recent experimental confirmation of its most counterintuitive prediction—that of *contextuality*, that is, the notion that the physical world depends on observation—contradict the classical view. According to quantum mechanics, the properties of tables and chairs exist *only insofar as they are observed*. Chapter 6 then shows how *essentially the same* idealist ontology discussed in Chapter 5 can make sense of this contextual quantum world. In doing so, it provides ontological underpinning for a parsimonious interpretation of quantum mechanics and opens up a new avenue of investigation for solving the so-called 'measurement problem.' More importantly, contextuality renders the mainstream physicalist metaphysics untenable, so Chapter 6 attempts to articulate a resolution to what is essentially a colossal—if seldom discussed—contradiction in our present-day understanding of reality.

Several objections are often raised against the idealist notion that all reality is mental or, more accurately, phenomenal (except for Chapter 9, wherein I define its usage differently, throughout this volume the word 'mental' is used as a synonym of 'phenomenal'). For instance, if there is no physically objective world outside mind, then reality is a kind of dream. How can we all be having the same dream, then? Moreover, if mind extends into the universe as a whole, why can't we mentally influence the laws of physics? Finally, if the brain does not generate the mind, how can physical intervention in the brain—in the form of trauma, psychoactive drugs, etc.—change our mental states?

Part III addresses these objections and many other subtler ones. Chapter 8 argues that they are implicitly based on logical errors, such as conflation, circular reasoning and even outright misunderstandings of the implications of idealism. One particularly important line of criticism is more thoroughly refuted in Chapter 9: critics point out that, if all reality is in consciousness, then there can't be such a thing as an *un*conscious mental process. Yet, many recent experimental results indicate that seemingly unconscious mental processes are abundant in the human psyche. Chapter 9 bites this bullet and argues that we have good reasons—theoretical, clinical and experimental—to believe that all these processes are, in fact, conscious at some level, despite appearances to the contrary.

All viable ontologies must be consistent with *all* reliable empirical evidence. As discussed in Chapters 6 and 15, this is *not* the case for physicalism: recent experimental results in quantum mechanics seem to directly contradict it. Nonetheless, Chapter 5 charitably ignores this physical line of evidence and argues that idealism is *still* superior to physicalism based solely on internal logic, parsimony and explanatory power. Because quantum mechanics often seems so abstract and removed from everyday reality, I did not want my argument in Chapter 5 to rest upon it.

However, overlooking evidence from neuroscience is much less defensible—if at all. After all, the relationship between mind and brain is of daily significance for everyone. The problem is that physicalism leaves this relationship largely *undefined*. It does not specify how the brain allegedly constitutes or generates the mind, so it can in principle accommodate *any* neuroscientific observation. Indeed, even physicalist and self-proclaimed skeptic Michael Shermer has gone on record admitting that "the neuroscience surrounding consciousness" is "nonfalsifiable" (2011). This means that physicalism manages to remain consistent with the evidence simply by *not* explaining the mind—primary fact of existence—to begin with. Its vague formulation prevents it from being pinned down by neuroscientific observations.

Be it as it may, we can still pose the following question: Which ontology—physicalism or idealism—makes *more sense* in view of the available neuroscientific evidence? In other words, is the evidence more consistent with what one would expect under physicalism or idealism? Part IV of this book makes the case that the observed correlations between brain activity and mental states are more consistent with idealism. In fact, some recent neuroscientific observations outright *contradict* physicalist expectations, whilst remaining elegantly in accord with idealist ones. Chapter 11 discusses a broad pattern of such observations. Chapter 12 then goes more in depth into one particular case: that of psychedelic trances. More than just showing how idealism can make more sense of psychedelic experiences, the chapter argues that physicalism—despite its vagueness—does have one *unavoidable* implication concerning the relationship between mind and brain. And this implication appears to have been experimentally contradicted by neuroimaging studies of psychedelic trances.

Having made explicit the thought-traps underlying today's academically fashionable ontologies (Part I), formulated a sound idealist alternative from both classical and quantum mechanical perspectives (Part II), refuted the main objections raised against this alternative (Part III) and then analyzed the neuroscientific evidence that seems to corroborate it (Part IV), in Part V I take a step back to contemplate how all this relates to us as individuals.

Indeed, if—as I hope to demonstrate—it is fairly simple to see that idealism is superior to physicalism, why has our mainstream cultural narrative been dominated by physicalism for at least well over a century now? To answer this question, Chapter 14 explores the implicit psychological motivations behind the adoption of physicalism. Moreover, if idealism is our best explanation for what is going on, what are its implications regarding the way we relate to life and the world? An attempt to tackle this question is made in Chapter 15.

Finally, the book closes with extensive additional commentary on the various ideas presented and an assessment of this work's place and role in our present cultural nexus.

An appendix reproduces an article on the implications of idealism regarding the after-death state, which didn't meet the criteria for inclusion in the main body of this work because it was an invited contribution to the journal wherein it first appeared.

Part I

What is wrong with the contemporary philosophical outlook

Contemporary methods employ predominantly dualistic procedures that do not extend beyond simple subject-object relationships; they limit our understanding to what is commensurate with the present Western mentality.
Jean Gebser: *The Ever-Present Origin.*

To learn more about mental aspects of the world ... we should try to discover 'manifest principles' that partially explain them, though their causes remain disconnected from what we take to be more fundamental aspects of science. The gap might have many reasons, among them, as has repeatedly been discovered, that the presumed reduction base was misconceived.
Noam Chomsky: *What Kind of Creatures Are We?*

Chapter 1

Preamble to Part I

A natural and perhaps even necessary first step in a book that aims to offer an alternative account of reality is to highlight what is wrong with the current approaches. After all, why bother with alternatives if the *status quo* is fine? As such, my intent in the next two chapters is not to gratuitously attack my peers in science and philosophy, but to highlight the need and secure the intellectual space for what is later argued in Part II.

The fact is that the mainstream physicalist ontology fails rather spectacularly to account for the most present and sole undeniable aspect of reality: the qualities of experience (see the "hard problem of consciousness" in Chalmers 2003). Physicalism is also arguably irreconcilable with results now emerging from physics laboratories around the world (e.g. Kim et al. 2000, Gröblacher et al. 2007, Romero et al. 2010, Lapkiewicz et al. 2011, Ma et al. 2013, Manning et al. 2015, Hensen et al. 2015, etc.), unless one takes so many liberties with the meaning of the word 'physicalism' that its spirit is outright contradicted. So both in terms of its explanatory power and its consistency with empirical observations, our mainstream view of the nature of reality is found wanting. Academically popular alternatives, such as bottom-up panpsychism, face many of the same empirical challenges, as well as analogous limitations in terms of explanatory power (see, for instance, the "subject combination problem" in Chalmers 2016).

Yet, my purpose with the next two chapters is not to compile a long and tedious list of the empirical and philosophical challenges faced by current ontologies. These challenges are well known and recognized in scientific and philosophical circles, there being no need to further stress them. What I want to attempt here is something more ambitious and—hopefully—ultimately more constructive: to point out the failures and internal contradictions *of the very thought processes* that led to these flawed ontologies in the first place. Only by understanding these failures and contradictions, as underlying causes of our present philosophical dilemmas, can we hope to reform our thinking and eventually solve the dilemmas.

In this context, Chapter 2 discusses what is perhaps the root of our contemporary philosophical ailment: the tendency to attempt to explain things by replacing concrete reality with abstractions. Such attempts consist by and large of mere word games, played in thought with a rich phantasmagoria of concepts, and represent perhaps the single greatest threat to our ability to remain grounded in reality in the 21st century. In a cultural environment that, because of the gap left open by the loss of religious myths, tacitly *expects* the latest scientific and philosophical theories to dazzle and boggle the mind (see Kastrup 2016a), scientists and philosophers alike seem ever more willing to lose themselves in a forest of abstractions highly prone to category mistakes.

This cultural legitimization of explanation by ungrounded abstraction is a hydra with many heads. Chapter 3 represents my attempt to identify these heads and diagnose the *specific* intellectual afflictions behind quandaries such as the "hard problem of consciousness" and the "subject combination problem." I hope to show that these quandaries are merely artifacts of unanchored thought, with no grounding in non-conceptual reality.

Later in the book, in Parts II to IV, I attempt to back up the legitimacy of the criticisms laid out in this Part I by offering an alternative way of thinking, as well as corresponding ontology, which overcome the intellectual afflictions alluded to above. As such, I hope to not only talk the talk, but also walk the walk. Insofar as I succeed in fixing the errors they point to, the criticisms in the next two chapters are given validation. May these criticisms thus be judged not by their incisiveness, but by my ability to demonstrate that a philosophical approach exists that does not fall prey to them.

Chapter 2

Conflating abstraction with empirical observation: The false mind-matter dichotomy

At the time of this writing, this article was scheduled to appear in *Constructivist Foundations*, ISSN 1782-348X, Vol. 13, No. 3, in July 2018. *Constructivist Foundations* is an interdisciplinary journal published by Alexander Riegler (Free University of Brussels) and thirty board members. It is indexed in Thomson Reuters's Arts & Humanities Citation Index (AHCI) and, in 2016, held the second highest ranking (Q2) in the Scimago Journal Rankings, a well-recognized measure of an academic journal's prestige.

2.1 Abstract

The alleged dichotomy between mind and matter is pervasive. Therefore, the attempt to explain matter in terms of mind (idealism) is often considered a mirror image of that of explaining mind in terms of matter (mainstream physicalism), in the sense of being structurally equivalent despite being reversely arranged. I argue that this is an error arising from language artifacts, for dichotomies must reside in the same level of abstraction. Because matter outside mind is not an empirical fact, but an explanatory model instead, the epistemic symmetry between the two is broken. Consequently, matter and mind cannot reside in the same level of abstraction. It becomes then clear that attempting to explain mind in terms of matter is epistemically more costly than attempting to explain matter in terms of mind. The qualities of experience are suggested to be not only epistemically, but also ontologically primary. The paper highlights the primacy of perceptual constructs over explanatory abstraction on both epistemic and ontic levels.

2.2 Introduction

The (unexamined) assumption that mind and matter are jointly exhaustive and mutually exclusive concepts is pervasive today. In other words, many people implicitly take every aspect of reality to be either mental (e.g. thoughts, emotions, hallucinations) or physical (e.g. tables and chairs), mentality and physicality being polar opposites in some sense. Originating with Descartes and Kant (Walls 2003: 130), this dichotomy has been firmly entrenched in Western thought since at least the early nineteenth century. Eminent scholarly publications of the time, such as *The British Cyclopædia of Natural History*, lay it out unambiguously: "as mind is the opposite of matter in definition, the perfection of its exercise must be the opposite of that of the exercise of matter" (Partington 1837: 161). From the early twentieth century onwards, more nuanced formulations of the dichotomy were proposed. Alfred North Whitehead (1947), for instance, considered mind and matter *co-dependent* opposites. Even Henri Bergson, whose conception of an *élan vital* was meant to dilute the Cartesian split, was careful not to completely eradicate the dichotomy (Catani 2013: 94).

Indeed, this trend towards more nuanced formulations endures to this day. Philosopher David Chalmers, for instance, wrote that the "failure of materialism leads to a kind of *dualism*: there are both physical and nonphysical [i.e. mental] features of the world" (1996: 124). He speaks of *property* dualism (Ibid.: 125) to distinguish it from the discredited *substance* dualism of Descartes. Nonetheless, the essence of the dichotomy persists intact. Public endorsements of property dualism by influential science spokespeople, such as neuroscientists Christof Koch (2012a: 152) and Sam Harris (2016), lend academic legitimacy to it. Harris, for instance, claims that mind and matter each represent "half of reality" (Ibid.), making the implicit assumption that they have comparable epistemic status (that is, that matter is as confidently knowable as mind). So pervasive is this assumption that it has become integral to our shared cultural intuitions.

Whilst a fundamental dichotomy between mind and matter is readily accepted by large segments of the population—perhaps for psychological reasons (Heflick et al. 2015)—in philosophical circles the corresponding dualism is properly regarded as unparsimonious. For this reason, philosophy has historically attempted to explain one member of the alleged dichotomy in terms of the other. The ontology of idealism, for instance, attempts to reduce "all sense data to mental contents" (Tarnas 2010: 335), whereas mainstream physicalism—perhaps better labeled as 'materialism,' but which I shall continue to refer to as 'mainstream physicalism' for the sake of consistency with some of the relevant literature—attempts to reduce all mental contents to material arrangements (Stoljar 2016). To be more specific, idealism entails that mind is nature's fundamental ontological ground, everything else being reducible to, or grounded in, mind, whereas mainstream physicalism posits that nature's fundamental ontological ground is matter outside and independent of mind, everything else being reducible to, or grounded in, matter.

The problem is that the ingrained cultural intuition that mind and matter have comparable epistemic status tends to creep—unexamined—even into philosophical thought, leading to the tacit conclusion that idealism and mainstream physicalism are mirror images of each other, in the sense of being structurally equivalent despite being reversely arranged. In the present essay, I contend that this tacit conclusion is false because it overlooks important epistemic considerations: we do *not*—and fundamentally *cannot*—know matter as confidently as we know mind. By incorrectly positing that idealism incurs epistemic cost comparable to that of mainstream physicalism in at least some important sense, the tacit conclusion undervalues idealism and overvalues physicalism. This confusion may be a key enabler of physicalism's success in underpinning our present-day mainstream worldview. Once the tacit conclusion is properly examined and rectified, as attempted in this essay, idealism may emerge as a more plausible ontology than mainstream physicalism at least in terms of its epistemic cost.

Like Gilbert Ryle (2009), I argue that mind and matter do *not* form a dichotomy. My argument, however, does not depend—as Ryle's controversially does (Webster 1995: 483)—on equating mind with behaviors. Indeed, Ryle attempts to refute the alleged dichotomy by effectively relegating mind to the status of mere illusion (Ibid.: 461). My argument, instead, rests on the notion that mind and matter are not epistemically symmetric—a concept I shall formally define in Section 2.5—as members of a dichotomy must be. I do not deny mind, because it is epistemically primary: all knowledge presupposes mind.

That the notion of physically objective matter—that is, matter outside and independent of mind—is now largely taken for granted suggests cultural acclimatization to what is in fact a mere hypothesis. After all, physically objective matter is not an observable fact, but a conceptual explanatory device *abstracted from* the patterns and regularities of observable facts—that is, an *explanatory abstraction* (Glasersfeld 1987; more on this in Section 2.4). Indeed, there seems to be a growing tendency in science today to mistake explanatory abstraction for what is available to us empirically. This has been extensively documented before, but mostly in regard to clearly speculative ideas such as superstring theory and multiverse cosmologies (Smolin 2007). When it comes to the everyday notion of physically objective matter, however, many fail to see the same conflation at work.

To illustrate and highlight the conflation with an admittedly extreme example, Section 2.3 briefly reviews the ontology of pancomputationalism, which posits ungrounded computation as the primary element of reality (Piccinini 2015). Indeed, the idea of replacing physicalism with ontic pancomputationalism should provide a visceral demonstration of the epistemic cost of substituting explanatory abstraction for observable facts. In this context, my suggestion is that an analogous epistemic disparity exists between idealism and mainstream physicalism. In other words, if one is convinced that ontic pancomputationalism is absurd in comparison to physicalism, then—and on the same basis—one has reason to question the plausibility of mainstream physicalism in comparison to idealism.

Section 2.4 then elaborates more systematically on the different planes of abstract explanation used in science and philosophy. It provides the basis for the refutation of the alleged dichotomy between mind and matter later carried out in Section 2.5, which forms the core of this essay. Finally, Section 2.6 sums it all up.

Before we start, however, some terminology clarifications are needed. Throughout this essay, I use the word 'mind' in the sense of phenomenal consciousness. Following Nagel's original definition of the latter (1974)—which has since been further popularized by Chalmers (1996, 2003)—I stipulate that, if there is anything it is like to be a certain entity, then the entity is minded. As such, mind—as the word is used here—is epistemically primary, an assertion further substantiated in Section 2.4. In this sense, mind does not necessarily entail higher-level functions such as metacognition—that is, the knowledge of one's knowledge (Schooler 2002: 340)—or even a conscious sense of self as distinct from the world. It necessarily entails only the presence of phenomenal properties, in that it is defined as the substrate or ground of experience. Moreover, insofar as what we call 'concreteness' is itself a phenomenal property associated with the degree of clarity or vividness of experience, mind is the sole ground of concreteness. Anything allegedly non-mental cannot, by definition, be concrete, but is abstract instead, in the sense of lacking phenomenal properties.

I am well aware that the word 'mind' is used in entirely different ways—often decoupled from experience—in other contexts, such as e.g. philosophy of biology (Godfrey-Smith 2014) and artificial intelligence (Franklin 1997). Yet, I believe the usage I am defining here is adequate for the context of the present paper. And given this usage, experience can be coherently regarded as an excitation of mind, whereas mind can be coherently regarded as the substrate or ground of experience.

2.3 The epistemic cost of explanation by abstraction

By postulating a material world outside mind and obeying laws of physics, physicalism can accommodate the patterns and regularities *of* perceptual experience. *But it fails to accommodate experience itself.* This is called the 'hard problem of consciousness' and there is now vast literature on it (e.g. Levine 1983, Rosenberg 2004: 13-30, Strawson et al. 2006: 2-30). In a nutshell, the qualities of experience are irreducible to the parameters of material arrangements — whatever the arrangement is — in the sense that it is impossible even in principle to deduce those qualities from these parameters (Chalmers 2003).

As I elaborate upon in Section 2.5, the 'hard problem' is not merely hard, but fundamentally insoluble, arising as it does from the very failure to distinguish explanatory abstraction from observable fact discussed in this paper. As such, it implies that we cannot, *even in principle*, explain mind in terms of matter. But because the contemporary cultural ethos entails the notion that mind and matter constitute a dichotomy, one may feel tempted to conclude that there should also be a symmetric 'hard problem *of matter*' — that is, that we should not, even in principle, be able to explain matter in terms of mind. The natural next step in this flawed line of reasoning is to look for more fundamental ontological ground preceding both mind and matter; a *third* substrate to which matter and mind could both be reduced.

A good example of this line of reasoning is brought by ontic pancomputationalism, which posits that ungrounded information processing is what makes up the universe at its most fundamental level (Fredkin 2003). As such, ontic pancomputationalism entails that computation precedes matter ontologically. But "if computations are not configurations of physical entities, the most obvious alternative is that computations are abstract, mathematical entities, like numbers and sets" (Piccinini 2015). According to ontic pancomputationalism, even mind itself— psyche, soul—is a derivative phenomenon of purely abstract information processing (Fredkin n.a.).

To gain a sense of the epistemic cost of this line of reasoning, consider the position of physicist Max Tegmark (2014: 254-270): according to him, "*protons, atoms, molecules, cells* and *stars*" are all redundant "baggage" (Ibid.: 255). Only the mathematical parameters used to describe the behavior of matter are real. In other words, Tegmark posits that reality consists purely of numbers—ungrounded information—but nothing to attach these numbers to. The universe supposedly is a "set of abstract entities with relations between them," which "can be described in a baggage-independent way" (Ibid.: 267). He attributes all ontological value to a description while—paradoxically— denying the existence of the very thing that is described in the first place.

Clearly, ontic pancomputationalism represents total commitment to abstract mathematical concepts as the foundation of reality. According to it, there are only numbers and sets. But what are numbers and sets without the mind or matter where they could reside? It is one thing to state in language that numbers and sets can exist without mind and matter, but it is an entirely other thing to explicitly and coherently conceive of what—if anything—this may mean. By way of analogy, it is possible to *write*—as Lewis Carroll did—that the Cheshire Cat's grin remains after the cat disappears, but it is an entirely other thing to conceive explicitly and coherently of what this means.

Ontic pancomputationalism appeals to ungrounded information—pure numbers, mathematical descriptions—as ontological primitive. But what exactly is information? Our intuitive understanding of the concept has been cogently captured and made explicit by Shannon (1948): information is given by state differences discernible in a system. As such, it is a property *of* a system—associated with the system's possible configurations—not an entity or ontological class unto itself. Under mainstream physicalism—that is, materialism—the system whose configurations constitute information is a material arrangement, such as a computer. Under idealism, it is mind, for experience entails different phenomenal states that can be qualitatively discerned from one another. Hence, information requires a mental or material substrate in order to be even conceived of explicitly and coherently. To say that information exists in and of itself is akin to speaking of spin without the top, of ripples without water, of a dance without the dancer, or of the Cheshire Cat's grin without the cat. It is a grammatically valid statement devoid of any semantic value; a language game less meaningful than fantasy, for internally consistent fantasy can at least be explicitly and coherently conceived of and, thereby, known as such. But in what way can we know information uncouched in mind or matter?

One assumes that serious proponents of ontic pancomputationalism are well aware of this line of criticism. How do they then reconcile their position with it? A passage by Luciano Floridi—well-known advocate of information as ontological primitive—may provide a clue. In a section titled "The nature of information," he states:

> Information is notoriously a polymorphic phenomenon and a polysemantic concept so, as an explicandum, it can be associated with several explanations, depending on the level of abstraction adopted and the cluster of requirements and desiderata orientating a theory. ... *Information remains an elusive concept.* (2008: 117, emphasis added.)

Such ambiguity lends ontic pancomputationalism a kind of conceptual fluidity that renders it impossible to pin down. After all, if the choice of ontological primitive is given by "an elusive concept," how can one definitely establish that the choice is wrong? In admitting the possibility that information may be "a network of logically interdependent but mutually irreducible concepts" (Ibid.: 120), Floridi seems to suggest even that such elusiveness may be unresolvable.

While vagueness may be defendable in regard to natural entities conceivably beyond the human ability to apprehend, it is at least difficult to justify when it comes to a *human concept* such as information. We *invented the concept*, so we either specify clearly what we mean by it or our conceptualization remains too ambiguous to be ontologically meaningful. In the latter case, there is literally *no sense* in attributing ontological value to information and, hence, ontic pancomputationalism is—once again—strictly meaningless.

Although ontic pancomputationalism is an admittedly extreme example, an analogous attempt to reduce concreteness— that is, the felt presence of conscious perception (Merleau-Ponty 1964)—to mere explanatory abstraction lies behind both mainstream physicalism and the alleged mind-matter dichotomy, as I shall argue in the next section. At the root of this concerning state of affairs is a generalized failure to recognize that every step of explanatory abstraction away from the concreteness of conscious perception implies a reduction in epistemic confidence: we do not know that abstract conceptual objects exist with the same level of confidence that we *do* know that our perceptions—whatever their source or underlying ontic nature may be—exist. I do not know that subatomic particles outside and independent of mind exist with the same level of confidence that I *do* know that the chair I am sitting on, which I am directly acquainted with through conscious perception, exists. Worse yet, with what confidence can we know that a loosely-defined, possibly incoherent concept such as ungrounded information lies at the foundation of reality? As such, steps of explanatory abstraction can only be justified if the observable facts cannot be explained *without* them, lest we conflate science and philosophy with meaningless language games. This is an important claim, so allow me to dwell on it a little longer before proceeding to the next section.

It could be argued that the existence of perceptual illusions indicates that conscious perception in fact entails *less* epistemic confidence than abstract formal systems. For instance, in the well-known 'checker shadow' illusion created by the Perceptual Science Group of the Massachusetts Institute of Technology (Adelson 1995), two identically-colored squares of a checkerboard are initially perceived to be of opposite colors because of the different contexts in which they are perceived. Should we then declare that conscious perception is fundamentally unreliable? Well, notice that *it is also conscious perception that eventually dispels the illusion*: by looking at one of the squares as it is moved to the other's context, one sees that it indeed has the same color as the other square. So even in the case of perceptual illusions, it is still direct, concrete experience that provides us with the epistemic confidence necessary to recognize the illusion for what it is.

Further supporting the claim that abstracting away from direct experience implies a reduction in epistemic confidence is the anti-realist view in philosophy of science. According to it, abstract theoretical entities—such as subatomic particles, invisible fields and any other postulated entity that escapes our ability to *directly* perceive—are but "convenient fictions, designed to help predict the behavior of things in the observable world" (Okasha 2002: 61; see also van Fraassen 1990). In other words, the best we can say about subatomic particles and other abstract entities is that the observable world behaves *as if* these abstract entities existed. This does not entail or imply that the entities *actually exist*, which we cannot be certain of either way (van Fraassen 1980). In this sense, explanatory abstraction again implies reduction in epistemic confidence, insofar as we do not know that subatomic particles and invisible fields exist with the same level of confidence that we do know that the world we consciously perceive exists.

2.4 Levels of explanatory abstraction

Like ontic pancomputationalism, mainstream physicalism is no stranger to the epistemic cost of explanatory abstraction: the existence of a material world outside and independent of mind is a theoretical inference arising from *interpretation* of sense perceptions within a framework of complex thought, not an observable empirical fact. After all, what we call the world is available to us solely as 'images' — defined here broadly, so to include any sensory modality — on the screen of perception, which is itself mental. Even physicist Andrei Linde, of cosmic inflation fame, acknowledged this:

> Let us remember that our knowledge of the world begins not with matter but with perceptions. I know for sure that my pain exists, my "green" exists, and my "sweet" exists ... everything else is a theory. Later we find out that our perceptions obey some laws, which can be most conveniently formulated if we assume that there is some underlying reality beyond our perceptions. This model of material world obeying laws of physics is so successful that soon we forget about our starting point and say that matter is the only reality, and perceptions are only helpful for its description. (1998: 12)

Now, we know that mind is capable to autonomously generate the imagery we associate with matter: dreams and hallucinations, for instance, are often qualitatively indistinguishable from the 'real world.' Therefore, the motivation for postulating an objective material world must go beyond the mere existence of this imagery. And indeed, what the notion of objective matter attempts to make sense of are certain *patterns and regularities observable in the imagery,* such as:

1. The correlations between observed brain activity and reported inner life (see e.g. Koch 2004 for a scientific take on the neural correlates of consciousness, but consider also the obvious effects of e.g. alcohol consumption and head trauma—both of which disrupt regular brain activity—on inner experience);
2. The fact that we all seem to inhabit the same world; and
3. The fact that the dynamics of this world unfold independently of our personal volition.

After all, if mind is not a product of objective arrangements of matter, how can there be such tight correlations between brain activity and experience? If the world is not made of matter outside our individual minds, how can we all share the same world beyond ourselves? If the world is not independent of mind, why can we not change the laws of nature simply by imagining them to be different? Clearly, thus, the non-mental world posited by physicalism is largely an attempt to make sense of these three basic observations. As such, it is an *explanatory abstraction,* not itself an observation. We conceptually *imagine* that there is a non-mental world underlying our perceptions — and in some sense isomorphic to these perceptions[1] — because doing so helps explain the basic observations. See Figure 2.1. Nonetheless, whatever ontological class is pointed to by this conceptual abstraction remains perforce epistemically inaccessible, a recognition already present in Kant's *Critique of Pure Reason.*

---

1 To say that *A* is isomorphic to *B* means that there is, in some sense, a correspondence of form between *A* and *B*.

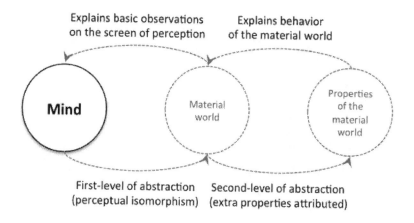

Figure 2.1: Levels of explanatory abstraction.

Explanatory abstraction does not stop at this first level. After imagining a non-mental world isomorphic to our perceptions, we are left with the task of explaining how and why this world behaves the way it does. Why do objects fall when dropped? Why does a piece of amber attract chaff when rubbed? How can certain metals magnetically attract other metals? To answer these questions, we must attribute to the material world certain properties that go beyond perceptual isomorphism. We say, for instance, that matter has the properties of mass, charge and spin. These properties constitute a second-level of explanatory abstraction beyond direct experience. See Figure 2.1 again.

Naturally, there can be even more levels of explanatory abstraction involved. Superstring theory, for instance, attempts to explain the properties of matter through the particular modes of vibration of imagined hyper-dimensional strings (Greene 2003). But the two levels illustrated in Figure 2.1 are sufficient for the discussion that follows.

The defining characteristic of explanation by abstraction is a progressive movement away from Husserl's "life-world" (1970), from the concreteness of direct experience. First, one posits a world devoid of qualities (Varela, Thompson & Rosch 1993) and, as such, devoid of concreteness too, for concreteness is a quality of experience. Then, one progressively loads this world with properties that entail no direct isomorphism to experience. For instance, we do not see electric charge or spin; we only see the behavior of matter that these abstract properties supposedly explain, such as attraction and repulsion. Similarly, we do not feel mass; we only feel the weight and inertia of objects, which the property of having mass supposedly explains (Okasha 2002: 58-76).

Because concreteness is the intuitive foundation of what we consider *real*, each step in this movement away from concreteness takes us farther from the only reality we actually know (Merleau-Ponty 1964). One may then become lost in a forest of intellectually appealing but ultimately arbitrary conceptualizations. This, again, is the epistemic cost of explanation by abstraction.

2.5 Dispelling the mind-matter dichotomy

By definition, the two members of a dichotomy are jointly exhaustive and mutually exclusive. Ontologically, this means that if one member is the case, then the other is necessarily *not* the case, and vice-versa. For instance, in the context of biological organisms, if life is *not* the case, then death is necessarily the case. In the context of a job application, if success is the case (i.e. the applicant gets the job), then failure is *not* the case. And so on. As such, a *single test* suffices to acquire knowledge about the ontological status of *both* members of a dichotomy. If I can perform a test to determine if a person is alive, then I will automatically know whether the person is dead, without having to test for death separately. If I can set a criterion for success, then this same criterion will automatically determine whether failure is the case, without my having to set a separate criterion for failure. And so on. I shall call this property of a dichotomy *epistemic symmetry*. When two concepts are epistemically symmetric, knowledge of one implies knowledge of the other.

Now notice that *epistemic symmetry can only hold for concepts residing in the same level of explanatory abstraction*. If they do not, then there necessarily is at least one extra inferential step necessary to know whether one of the concepts obtains. This breaks the symmetry, for then we cannot acquire knowledge of the ontological status of both concepts with a single test.

Here is an example: the presence of a negative feeling can be tested for directly through introspection—thus entailing no inferential steps—whereas testing for the presence of a positive electric charge requires an inference by observation of the associated behavior of matter. Because of this need for an extra inferential step, knowing the negative feeling cannot imply knowledge of the positive electric charge. The negative feeling and the positive electric charge are not, therefore, epistemically symmetric and cannot constitute a dichotomy.

Conversely, positive and negative electric charges are both properties of matter, residing in the second level of explanatory abstraction illustrated in Figure 2.1. As such, they are epistemically symmetric and can constitute a dichotomy. As a matter of fact, every level of explanatory abstraction can encompass dichotomies. For instance, the size of material objects is isomorphic to perceptual qualities: we can subjectively test whether an object is big or small in relation to another object. As such, bigness and smallness both reside in the first level of explanatory abstraction and are epistemically symmetric; they can constitute a dichotomy. See Figure 2.2.

But—and here is the key point—*mind and matter do not reside in the same level of explanatory abstraction*. In fact, mind—as defined in Section 2.2—is the ground within which, and out of which, abstractions are made. Matter, in turn, is an abstraction *of* mind (see Figure 2.1 again). This breaks the epistemic symmetry between them: we do not know matter in the same way that we know mind, for—as cogently argued by Linde in the earlier quote—matter is an inference and mind a given. Consequently,

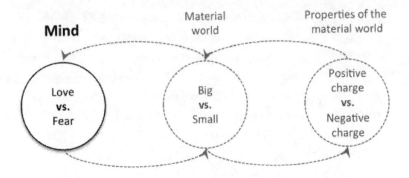

Figure 2.2: Dichotomies in their respective levels of explanatory abstraction.

although mind can encompass polar opposites—such as the feelings of love and fear in the context of a situation wherein someone feels passionate about a particular aspect of someone else (assuming that other passions, such as hate, which is arguably a form of fear, are in fact particular instances of love or fear)—it cannot itself be the polar opposite of matter or matter's properties. It follows that we have no reason to conclude that reducing matter to mind is as challenging as reducing mind to matter, and there is thus no substantiation for a 'hard problem of mind.' Stronger still, insofar as what we call 'matter' can be parsimoniously construed as phenomenal patterns of excitation of mind, matter is on an epistemic par with mind and can in principle be reduced to the latter, for both already reside in the same ontological domain. This move takes mind itself to be an ontological primitive and eliminates any conceivable 'hard problem of mind,' since mind now does not need to be reduced.

The notion of a dichotomy between mind and matter arises from language. In order to speak of the substrate of experience we must give it a name, such as 'mind' or 'consciousness,' thereby linguistically objectifying what is in fact the subject. Then, we conflate language with reality, implicitly assuming that mind is an object just as matter allegedly is. We forget that, in fact, there is no epistemic symmetry between the two.

Indeed, because the concept of mind-independent matter, as an explanatory abstraction, arises *in mind*, as an 'excitation' *of* mind, to say that mind and matter constitute a dichotomy is akin to saying that ripples and water constitute a dichotomy. Dichotomies can exist only between different kinds of ripples — say, those that flow mostly to the right versus those that flow mostly to the left — not between ripples and the substrate where they ripple. Mind is the substrate of the explanatory abstraction we call matter, so when we speak of a mind-matter dichotomy we incur in a fundamental "category mistake," as Ryle (2009) put it. *However, contrary to what Ryle suggests, it is matter that is the abstraction, not mind.*

The notion that idealism and mainstream physicalism are mirror images of each other arises from a failure to grasp this point. Lucid contemplation of these ontologies shows that idealism attempts to reduce an explanatory abstraction (physically objective matter) to that which articulates and hosts the abstraction in the first place (mind). This is *prima facie* eminently reasonable. Mainstream physicalism, in turn, attempts to reduce mind to mind's own explanatory abstractions, an obvious paradox that constitutes the crux of the 'hard problem.'

There would be no 'hard problem' if one did not conflate explanatory abstractions with concrete ontological primitives; if one did not attempt to paradoxically reduce mind to abstractions *of* mind. The 'hard problem' is not an empirical fact but the salient result of internal contradictions in a logico-conceptual schema; contradictions that I hope to have helped make explicit with the present paper.

Naturally, circumventing the 'hard problem' in the way suggested above ultimately forces us to make do with mind alone as ontological primitive and thereby entertain some form of idealism — more specifically, a form of idealism wherein mind is the experientially given ground of reality, whose manifestations comprise the concrete phenomenality you and I undergo in everyday life. And whereas idealism in the West has had its heyday in the eighteenth (e.g. Berkeley) and early nineteenth (e.g. Hegel) centuries, it is now enjoying renewed interest (Chalmers forthcoming) for having been updated and revitalized with compelling new formulations (e.g. Kastrup 2017b[2] and 2017e,[3] Yetter-Chappell forthcoming, as well as Fields et al. 2017, insofar as the latter can be construed as a form of idealism). These are sometimes proposed under new names, such as 'cosmopsychism' (e.g. Shani 2015, Nagasawa & Wager 2016), which, as the name suggests, posits that the cosmos as a whole is essentially phenomenal. Even 'radical constructivism' can be construed as a form of idealism, insofar as its claims are not merely epistemic, but ontic: "Radical constructivism … develops a theory of knowledge in which knowledge does not reflect an 'objective' ontological reality, but *exclusively* an ordering and organization of a world *constituted* by our experience" (Glasersfeld 1987: 199, emphasis added). Finally, the strongest objections usually leveraged against idealism have recently also been tackled (Kastrup 2017c[4]).

Having said all this, it should be noted that, in and of itself, the argument provided in this paper, despite being supportive of idealism, does not necessarily *imply* idealism. I have focused on epistemic cost considerations and did not show whether or how idealism can account for all relevant facts of nature. Indeed, an articulation of an idealist ontology is not in the scope of this paper.[5] But if it is demonstrated—as some of the papers cited above claim to do—that idealism *can* account for all facts that mainstream physicalism allegedly accounts for, then epistemic

---

2 This article can be found in Chapter 5 of the present volume.

3 This article can be found in Chapter 6 of the present volume.

4 This article can be found in Chapter 8 of the present volume.

5 For such an articulation, see Part II of the present volume.

cost considerations certainly tilt the balance in favor of idealism, due to the latter's lack of reliance on inflationary, epistemically unreliable, paradoxical abstractions. As such, the core claim of this essay is not as much the validity of idealism as that physically objective matter is a doubtful *cognitive construct*, in the strict constructivist sense: insofar as we believe to see matter outside and independent of mind when we look at the world around ourselves, we are in fact conflating a rational-linguistic construction with reality itself.

2.6 Conclusions

The pervasive but unexamined assumption that mind and matter constitute a dichotomy is an error arising from language artifacts. Members of true dichotomies must be epistemically symmetric and, therefore, reside in the same level of abstraction. Physically objective matter—as an explanatory model—is an abstraction *of* mind. We do not *know* matter in the same way that we know mind, for matter is an inference and mind a given. This breaks the epistemic symmetry between the two and implies that mainstream physicalism and idealism cannot be mirror images of one another.

Failure to recognize that different levels of epistemic confidence are intrinsic to different levels of explanatory abstraction lies at the root not only of the false mind-matter dichotomy, but also of attempts to make sense of the world through increasingly ungrounded explanatory abstractions. Lest we conflate science and philosophy with hollow language games, we must never lose sight of the difference between an abstract inference and a directly observable fact. Keeping this distinction in mind allows us to construct useful predictive models of nature's *behavior*—which ultimately is what science is meant to do—without restrictive and ultimately fallacious inferences about what nature *is*. This, in turn, liberates us from thought artifacts such as the 'hard problem of consciousness' and opens up whole new avenues for making sense of self and world.

Chapter 3

The quest to solve problems that don't exist: Thought artifacts in contemporary ontology

This article first appeared in *Studia Humana*, ISSN: 2299-0518, Vol. 6, No. 4, pp. 45-51, on 16 October 2017. *Studia Humana* is published by *De Gruyter*, a large German publisher of scholarly literature whose roots go back as far as 1749. Today *De Gruyter* claims to be the world's third largest open-access academic publisher.[1]

3.1 Abstract

Questions about the nature of reality and consciousness remain unresolved in philosophy today, but not for lack of hypotheses. Ontologies as varied as physicalism, microexperientialism and cosmopsychism enrich the philosophical menu. Each of these ontologies faces a seemingly fundamental problem: under physicalism, for instance, we have the 'hard problem of consciousness,' whereas under microexperientialism we have the 'subject combination problem.' I argue that these problems are thought artifacts, having no grounding in empirical reality. In a manner akin to semantic paradoxes, they exist only in the internal logico-conceptual structure of their respective ontologies.

3.2 Introduction

While advances in technology—enabled by the predictive

---

1 See: https://www.degruyter.com/dg/page/79/eine-kurze-geschichte-des-verlags (accessed 27 July 2017).

models of science—have influenced early 21st century culture more than anything else, questions of ontology loom large in the contemporary psyche: What is the nature of reality? What is the essence of phenomenal consciousness and how does it relate to matter? Our tentative answers to these questions color—if not outright determine—our view of life's meaning, thereby underlying every aspect of our existence.

Philosophy has not been idle in the face of demand for a menu of hypotheses in this regard. The mainstream physicalist ontology, for instance, posits that reality is constituted by irreducible physical entities—which Strawson has called 'ultimates' (Strawson et al. 2006: 9)—outside and independent of phenomenality. According to physicalism, these ultimates, in and of themselves, do not instantiate phenomenal properties. In other words, there is nothing it is like to be an ultimate, phenomenality somehow emerging only at the level of complex arrangements of ultimates. As such, under physicalism phenomenality is not fundamental, but instead reducible to physical parameters of arrangements of ultimates.

What I shall call 'microexperientialism,' in turn, posits that there is already something it is like to be at least some ultimates, combinations of these experiencing ultimates somehow leading to more complex experience (Strawson et al. 2006: 24-29). As such, under microexperientialism phenomenality is seen as an irreducible aspect of at least some ultimates. The ontology of panexperientialism (Griffin 1998: 77-116, Rosenberg 2004: 91-103, Skrbina 2007: 21-22) is analogous to microexperientialism, except in that the former entails the stronger claim that all ultimates instantiate phenomenal properties.

Micropsychism (Strawson et al. 2006: 24-29) and panpsychism (Skrbina 2007: 15-22) are analogous—maybe even identical—to microexperientialism and panexperientialism, respectively, except perhaps in that some formulations of the former admit cognition—a more complex form of phenomenality—already at the level of ultimates, as an irreducible aspect of these ultimates. Among microexperientialism, panexperientialism, micropsychism and panpsychism, microexperientialism makes the narrowest claim and, therefore, is the most generic. In a strong sense, panexperientialism, micropsychism and panpsychism are variations or extensions of microexperientialism, the latter being the canonical basis of all four ontologies. Therefore, I shall henceforth speak only of microexperientialism.[2]

Whereas microexperientialism entails that bottom-up combinations of simple subjects give rise to more complex ones, such as human beings, cosmopsychism (Nagasawa & Wager 2016, Shani 2015) takes the opposite route: according to it, the cosmos as a whole is conscious, individual psyches arising from top-down discontinuity in the integration of the contents of cosmic consciousness. Cosmopsychism can also be interpreted so as to include the further claim that, in addition to being conscious, the cosmos has a facet irreducible to phenomenal properties: the physical universe we can measure. This implies a form of dual-aspect monism, *a la* Spinoza (Skrbina 2007: 88), so I shall call this interpretation 'dual-aspect cosmopsychism.' Under dual-aspect cosmopsychism, the cosmos as a whole *bears* phenomenality, but is not *constituted by* phenomenality. In other words, the cosmos is supposedly *conscious*, but not *in consciousness*.

My goal with this brief essay is to show that the thought processes underlying many of these ontologies are flawed, for being based on unexamined assumptions and unwarranted logical bridges. Once this is lucidly understood, some of the most important open questions associated with these ontologies — which contemporary philosophers see as their job to answer — are exposed as artifacts. Indeed, it is my contention that some of

---

2 Elsewhere in this book I use the label 'bottom-up panpsychism' to refer generically to microexperientialism, panexperientialism, micropsychism or panpsychism, without distinguishing between them.

the key problems of ontology that contemporary philosophers have been grappling with do not actually exist. The next sections will elaborate upon this claim.

Anticipating a point that is bound to be raised, I acknowledge that offering a coherent alternative to the ontologies I am about to criticize is important for the completeness of my argument. And as attentive readers will notice, only idealist ontologies—those entailing that all existence is essentially phenomenal—are left unscathed by the criticisms in this paper. For this reason, I have extensively elaborated on a formulation of idealism elsewhere (Kastrup 2017b[3]) and also rebutted many objections to it (Kastrup 2017c[4]). Here, however, I shall limit myself to deconstructing the rationale behind the mainstream physicalist ontology and two of its more recent alternatives. Readers interested in my formulation of idealism are referred to the works cited above.

3.3 Thought artifacts in physicalism

As discussed in the previous section, physicalism entails the existence of a world outside and independent of consciousness, which I shall henceforth refer to as the 'objective physical world.' This postulate seems to be self-evident from the perspective of modern and post-modern culture, yet it is merely a theoretical *inference* arising from interpretation of sense perceptions. After all, what we call the world is available to us solely as 'images'—defined here broadly, so to include any sensory modality—on the screen of perception, which is itself in consciousness. (To avoid possible misinterpretations, notice that my point here is agnostic of whether these perceptual images are a valid given—in the sense of being both epistemically independent and efficacious (Sellars 1997)—or not. My point is that, in either case, *the objective physical world is surely not a given.*)

---

3 This article can be found in Chapter 5 of the present volume.
4 This article can be found in Chapter 8 of the present volume.

Stanford physicist Prof. Andrei Linde perhaps explained best the inferential nature of the objective physical world:

> Let us remember that our knowledge of the world begins not with matter but with perceptions. I know for sure that my pain exists, my "green" exists, and my "sweet" exists. I do not need any proof of their existence, because these events are a part of me; everything else is a theory. Later we find out that our perceptions obey some laws, which can be most conveniently formulated if we assume that there is some underlying reality beyond our perceptions. This model of material world obeying laws of physics is so successful that soon we forget about our starting point and say that matter is the only reality, and perceptions are only helpful for its description. This assumption is almost as natural (and maybe as false) as our previous assumption that space is only a mathematical tool for the description of matter. But in fact we are substituting reality of our feelings by a successfully working theory of an independently existing material world. And the theory is so successful that we almost never think about its limitations until we must address some really deep issues, which do not fit into our model of reality. (1998: 12)

Now, we know that consciousness is perfectly capable to autonomously generate the imagery we associate with physicality: dreams and hallucinations, for instance, are often qualitatively indistinguishable from the 'real world.' Therefore, the motivation for positing the existence of an objective physical world must go beyond the mere existence of this imagery. And indeed, what physicalism attempts to make sense of are certain basic facts observable *in* the imagery, such as:

1.   The correlations between observed brain activity and reported inner life (cf. Koch 2004);

2. The fact that we all seem to inhabit the same world; and
3. The fact that the dynamics of this world unfold independently of personal volition.

After all, if consciousness isn't a product of objective arrangements of physical elements, how can there be such tight correlations between brain activity and experience? If the world isn't made of physical elements outside our individual psyches, how can we all inhabit the same world beyond ourselves? If the world isn't independent of consciousness, why can't we change the laws of nature simply by imagining them to be different? Clearly, thus, the objective physical world posited by physicalism is an attempt to make sense of these three basic facts. As such, it is an *explanatory model*, not itself an observation. We *imagine* that there is an abstract physical world underlying our perceptions—and in some sense isomorphic to these perceptions[5]—because doing so helps explain the basic facts.

Conjuring up an objective physical world to make sense of observations would—at least in principle—be legitimate if it didn't create an insoluble problem known as the 'hard problem of consciousness' (Chalmers 2003, Levine 1983). Indeed, one of physicalism's key tenets is that consciousness itself must be reducible to arrangements of objective physical elements. The problem, of course, is that it is impossible to conceive of how or why any particular structural or functional arrangement of physical elements would constitute or generate experience (Rosenberg 2004: 13-30, Strawson et al. 2006: 2-30). The qualities of experience are irreducible to the observable parameters of physical arrangements—whatever the arrangement is—in the sense that it is impossible to deduce those qualities—even *in principle*—from these parameters (Chalmers 2003). There is

---

5 To say that *A* is isomorphic to *B* means that there is, in some sense, a correspondence of form between *A* and *B*.

nothing about the momentum, mass, charge or spin of physical particles, or their relative positions and interactions with one another, in terms of which we could deduce the greenness of grass, the sweetness of honey, the warmth of love, or the bitterness of disappointment. As long as they fit with the observed correlations between neural activity and reported experience, mappings between these two domains are entirely arbitrary: in principle, it is as (in)valid to state that spin up generates the feeling of coldness and spin down that of warmth as it is to say the exact opposite. There is nothing intrinsic about spin—or about any other parameter of physical elements or arrangements thereof—that would allow us to make the distinction.

For this reason, neuroscience finds itself positing a slew of conflicting speculative theories about the neural constitutors or generators of experience, varying from information integration across vast networks of neurons (Tononi 2004) to microscopic intra-neural dynamics (Hameroff 2006). Indeed, as skeptic Michael Shermer wrote, "the neuroscience surrounding consciousness" is "nonfalsifiable" (2011). Such nonfalsifiability derives from the fact that the logical bridge between the felt qualities of experience and the configurations of an abstract world beyond experience is arbitrary.

Let us take a step back and unpack the thought process that brought us to this dilemma: first, the consciousness of a physicalist wove the conceptual notion that some patterns of its own dynamics—namely, those of sense perception—must somehow exist outside itself; then, the consciousness of the physicalist tried to project its own essence onto these patterns. The glaring artifact of thought here becomes apparent with an analogy: imagine a painter who, having painted a self-portrait, points at it and declares himself to *be* the portrait. This, in essence, is what physicalism does. The consciousness of the physicalist conceptualizes self-portraits within itself. Sometimes these self-portraits take the form of electrical impulses and neurotransmitter releases in the brain (Koch 2004). Other times, they take the shape of quantum transitions or potentials (Tarlaci & Pregnolato 2016). Whatever the case, the physicalist's consciousness always points to a conceptual entity it creates within itself and then declares itself to *be* this entity. It dismisses its own primary, first-person point of view in favor of an abstract third-person perspective. Consider Daniel Dennett's words: "The way to answer these 'first-person point of view' stumpers is *to ignore the first-person point of view* and examine what can be learned from the third-person point of view" (1991: 336, emphasis added). The contempt for direct experience, primary datum of existence, is palpable here.

This arbitrary dislocation of epistemic primacy from direct experience to explanatory abstraction is what conjures up the 'hard problem.' If we didn't insist that direct experience must somehow be constituted or generated by 'something beyond' direct experience, there would be no problem. And since this 'something beyond' is a conceptual invention derived from an explanatory model, the 'hard problem' itself is a conceptual invention.

The issue here is that the invention forces the physicalist into the impossible position of *having to reduce consciousness to consciousness's own abstractions*. This is as absurd as trying to reduce a painter to his paintings; cause to its effects. As such, the 'hard problem' is akin to a semantic paradox: the difficulty behind it is grounded not in empirical reality, but in its internal logico-conceptual structure.

For as long as they fail to remain alert to the fact that an objective physical world outside consciousness is a conceptual creation of consciousness itself, physicalists will continue to struggle with an insoluble problem. Indeed, the fundamental insolubility of the problem is itself a glaring hint that something has gone wrong in the underlying thought processes that led to it in the first place.

3.4 Thought artifacts in microexperientialism

As we have seen, microexperientialism posits that entities as small as subatomic particles are experiencing subjects in their own merit. Microexperientialists imagine that the unitary subjectivity of more complex experiencing subjects, such as human beings, arises from *bottom-up combination* of countless simpler subjects. This circumvents the 'hard problem' by positing that consciousness is a fundamental, irreducible property of ultimates and, as such, does not need to be explained in terms of anything else.

However, another problem immediately arises: the combination of subjects is an unexplainable process, perhaps incoherent (Coleman 2014). It is just as hard as the 'hard problem' itself (Goff 2009). We cannot coherently explain how or why any physical action—such as bringing two subatomic particles close together or having them interact in some way—would cause the unification of their subjective points of view, as required by microexperientialism. This is known in contemporary philosophy as the 'subject combination problem' (Chalmers 2016). And, just like the 'hard problem,' it is an artifact of thought.

Indeed, the motivation for microexperientialism is that subatomic particles are the discernible 'pixels' of the empirical world we perceive around ourselves.[6] But to imagine, for this reason, that the subjectivity of living beings is composed of myriad subatomic-level subjects makes a rather simple mistake: it attributes to *that which experiences* a structure discernible only *in the experience itself.*

Let us unpack this. The notion of fundamental subatomic particles—ultimates—arises from experiments whose outcomes are accessible to us only in the form of perception (even

---

6 That is, they are the elementary, indivisible building blocks of the images on the screen of perception, insofar as we can discern with the aid of instrumentation.

when delicate instrumentation is used, the output of this instrumentation is only available to us as perception). Such experiments show that the images we experience on the screen of perception can be divided up into ever-smaller elements, until we reach a limit. At this limit, we find the smallest discernible components of the images, which are thus akin to pixels. As such, ultimates are the 'pixels' of *experience*, not necessarily of the experienc*er*. The latter does not follow from the former.

Even the fact that human bodies are made of subatomic particles says nothing about the structure of the experienc*er*: what we call a human body is itself an image on the screen of perception, and so will necessarily be 'pixelated' insofar as it is perceived. Such pixelation reflects the idiosyncrasies *of the screen of perception*, not necessarily the structure of the human subject itself. As an analogy, the pixelated image of a person on a television screen reflects the idiosyncrasies *of the television screen*; it doesn't mean that the person itself is made up of pixels.

To conclude that a living subject—that is, the consciousness of a living being—is made up of a combination of lower-level inanimate subjects requires an extra logical step for which, unless we beg the question of ontology, there is no justification. It is analogous to saying, for instance, that water is made of ripples simply because one can discern individual ripples in water. Obviously, individual ripples make up the structure of the *movements* of water, not of water itself. Analogously, subatomic particles are the 'pixels' of the observable 'movements' of consciousness, not necessarily the building blocks of consciousness itself. We have just as much reason to conclude that our subjectivity is composed of myriad subatomic-level subjects as to conclude that water is made of ripples.

Clearly, thus, the 'combination problem' of micro-experientialism is an artifact of a fallacious logical bridge. Just like the 'hard problem' faced by physicalism, it is not grounded in empirical reality, but in the internal logico-conceptual structure of microexperientialism itself.

3.5 Thought artifacts in dual-aspect cosmopsychism

Dual-aspect cosmopsychism is the least problematic ontology among the three criticized in this brief essay. By positing that the cosmos as a whole is conscious, the associated cosmic consciousness being an irreducible aspect of reality, it circumvents both the 'hard problem' and the 'combination problem.' One might then be tempted to conclude that a third, equivalent problem must be incurred, which we might call the 'decomposition problem': How does one cosmic consciousness apparently break up into myriad individual psyches, such as yours and mine? This, however, is actually not a fundamental problem, for "a disruption of and/or discontinuity in the normal integration of consciousness" (Black & Grant: 191) that can account for the *appearance* of decomposition is well known and understood today, under the label of "dissociation" (American Psychiatric Association 2013).

So what is the thought artifact behind dual-aspect cosmopsychism then? It is the redundant and inflationary postulate that the cosmos as a whole is a *"bearer* of consciousness" (Shani 2015: 408, emphasis added), as opposed to being *constituted by* consciousness. For the cosmos to *bear* consciousness there must be something to it—some aspect of it—beyond consciousness itself, which can in turn carry consciousness. Otherwise, what sense is there in saying that consciousness bears consciousness? This postulate of dual-aspect cosmopsychism may be an unexamined concession to the reigning physicalist view that there exists something beyond phenomenality. By accommodating this view, dual-aspect cosmopsychism certainly becomes more digestible under the contemporary zeitgeist. However, the key challenge incumbent upon cosmopsychism is to explain how a unitary cosmic consciousness can give rise to apparently distinct individual psyches. The idea of a physically objective facet of the cosmos is not necessary or helpful for tackling and overcoming such a challenge (cf. Nagasawa & Wager 2016, Shani 2015). Therefore, by accommodating the physicalist view that there exists something beyond phenomenality, dual-aspect cosmopsychism also ends up incorporating a redundant and inflationary postulate.

If the notion of an objective physical world is left out of cosmopsychism, the latter boils down to idealism: the view that the cosmos as a whole is *in consciousness*—as opposed to being *conscious*—and that individual psyches arise from a process of top-down dissociation in cosmic consciousness (Kastrup 2017b[7]). Although idealism faces challenges regarding its explanatory power—that is, its ability to make sense of the facts that we all seem to share the same world outside the control of our volition, that physical interference with the brain clearly affects inner experience, etc.—it does not fall victim to any of the artifacts of thought discussed in this essay.

3.6 Conclusions

The key philosophical problems faced by today's most popular ontologies—such as the 'hard problem of consciousness' faced by physicalism and the 'subject combination problem' faced by microexperientialism—are artifacts of unexamined assumptions and fallacious logical bridges inherent to their respective ontologies, having no grounding in empirical reality. In a manner akin to semantic paradoxes, they exist only in the internal logico-conceptual structure of these ontologies. The sooner philosophers become lucid of this fact, the sooner philosophical thought can move towards more constructive avenues of inquiry.

---

7 Again, this article can be found in Chapter 5 of the present volume.

Part II

An idealist ontology

We posit the existence of stimuli to explain our perceptions of the world, and we posit their immutability to avoid both individual and social solipsism. About neither posit have I the slightest reservation. But our world is populated in the first instance not by stimuli but by the objects of our sensations.
Thomas Kuhn: *The Structure of Scientific Revolutions.*

It is to be suspected that our division into material versus mental, that which is observable from the outside versus that which is perceivable from the inside, is only a subjectively valid separation, only a limited polarization that our structure of consciousness imposes on us but that actually does not correspond to the wholeness of reality. In fact it is rather to be suspected that these two poles actually constitute a unitary reality.
Marie-Louise von Franz: *Psyche and Matter.*

Chapter 4

Preamble to Part II

The two chapters that follow comprise the core of this work. Based on careful consideration of the available empirical evidence and guided by parsimony and logical consistency, they argue that the best explanation for the facts of nature entails that these facts are essentially *phenomenal*. The dynamics of matter in the inanimate universe is simply the extrinsic appearance of *im*personal mental processes, in the same way that human brain activity is the extrinsic appearance of *personal* mental processes. As such, these chapters articulate and defend a present-day form of the ontology of *idealism*, according to which all existence consists solely of *ideas*: thoughts, emotions, perceptions, intuitions, imagination, etc.

The analogy above is the key to a *felt*—as opposed to merely conceptual—understanding of the next two chapters, for it takes to heart what nature itself has been making patently clear: *so-called 'material' brain activity is how a person's conscious inner life*—her thoughts, feelings, fantasies, beliefs, etc.—*appears to other people*. This, as stated, is a fact, not a theoretical inference. Nature is thus unequivocally telling us not only that there is something conscious inner life *looks like* from a second-person perspective, but that this 'something' takes the form of what we call 'matter' (the brain, after all, is made of matter). Any further conclusion—such as that matter is independent of mind and, in turn, somehow generates the qualities of experience when arranged in certain ways—is already the outcome of theory, not an observable fact. This book seeks to look at nature without theoretical preconceptions: if the matter in a working brain is the extrinsic appearance of conscious inner life, *then*—at least in principle—*so should the matter in the inanimate universe as a whole be*. After all—and again short of theoretical assumptions—why should matter be one thing when constituting a living brain and then something else when constituting the inanimate universe as a whole? When we contemplate the large-scale structure and dynamics of the cosmos, we must be contemplating the extrinsic appearance of universal *conscious* inner life. As such, when impartially observed and pondered, nature renders the ideas presented in the next two chapters entirely commonsensical, as opposed to a challenge to commonsense.

Nonetheless, to most people today the idealist view that nature is entirely mental may sound exceedingly counterintuitive. The world is not only concrete and enduring—as opposed to vague and ephemeral, such as the imagination—but also clearly independent of our personal volition. We can't walk through walls merely by wishing to do so. Moreover, we seem to *inhabit* a shared world, as opposed to hosting it in our psyche. The argument in the next two chapters reconciles these facts with idealism.

Before we begin, however, it is important to keep in mind the distinction between idealism and *solipsism*. According to solipsism, the world is your *individual* dream. The whole of existence unfolds in your individual psyche alone. All other seemingly conscious creatures are merely figments of your imagination; there is allegedly nothing it is like to be them.

This is *not* what idealism posits. According to idealism, the whole universe is in mind, *but not in your individual psyche alone*, for mind extends far beyond the boundaries of personal introspection. The outside world is indeed outside your individual mentation, just not outside mind as an ontological class. Idealism grants that other living organisms are truly conscious—that is, that there is something it is like to be them—and their appearances and behaviors aren't merely figments of your personal imagination. As such, idealism is different from solipsism and shouldn't be confused with it as you make your way through the next chapters.

Chapter 5 explains our classical world under idealism. The goal is to show that, even if there were no such things as quantum mechanics and its counterintuitive implications, the notion of a mental universe would *still* be the most parsimonious and powerful explanation for our daily experiences. The chapter argues that existence consists of patterns of self-excitation of one universal mind. We and other living organisms are dissociated alters of this universal mind, akin to the multiple disjoint personalities of a person with dissociative identity disorder. The inanimate universe we see around us is the extrinsic appearance of mentation in the segment of universal mind that is not comprised in any alter, which I shall call 'mind-at-large.' So the inanimate universe is indeed outside our individual psyches — that is, outside our respective alters — but still inside universal mind. We seem to inhabit the same shared world because we are all immersed in, and surrounded by, the ideas of mind-at-large.

Then, Chapter 6 bites the bullet of quantum weirdness by tackling what is technically called 'contextuality.' Basically, contextuality means that the properties of the physical world — e.g. the position and momentum of objects — do *not* exist independently of observation. The physical world we perceive isn't merely discovered by observation, but *created* by it. Weird as this may sound, contextuality is predicted by quantum theory and many recent experiments have corroborated it. Chapter 6 shows how idealism can make sense of all this *without* solipsist assumptions.

So while Chapter 5 explains a classical, *non*-contextual world under idealism, Chapter 6 explains a quantum, *contextual* world. These worlds are so different that one might expect each to require an entirely distinct ontological framework. Yet, such is not the case. With essentially the same ontology developed in Chapter 5, Chapter 6 makes sense of quantum weirdness: what we call the 'physical world' arises from an *interaction* — an interference pattern—between the internal *mental* state of our alter and the external *mental* state of mind-at-large. An observation *is* this interaction across dissociated mental domains, which explains contextuality.

Notice thus that the ontology in Chapter 5, despite being meant to explain the classical world of everyday experience, has inherent features that allow it to elegantly accommodate and make sense of contextuality. Reconciling classical and quantum worlds in this seamless manner is—I hope—also a key contribution of Part II.

Although contextuality—with its experimental confirmation—is seldom discussed outside the small and highly specialized community of foundations of physics, it renders untenable the naïve-realist notion that the physical world we perceive around ourselves exists autonomously. *We know, both theoretically and experimentally, that such is not the case.* Nonetheless, for whatever reason, this knowledge hasn't percolated through society. As a matter of fact, it hasn't percolated even through the broader scientific and philosophical communities, which largely continue to operate under a view of reality known to be false.

As such, idealism isn't just a *better* explanation for the world we perceive around ourselves; it is possibly the *only* viable explanation, insofar as the known alternatives—variations of physicalism and bottom-up panpsychism, as well as some interpretations of cosmopsychism—are untenable in view of contextuality. This observation is sobering and has been a key motivation for the publication of this volume.

Moreover, the brief discussion in Chapter 6 about experimental results corroborating contextuality—which are later explained much more extensively, in layman's terms meant for the non-physicist reader, in Chapter 15—constitutes a compelling *empirical* case for idealism. I have not included it in Part IV of this book partly because it is intrinsically intertwined with the argument in Chapter 6. Consequently, Part IV focuses only on the *neuroscientific* line of empirical evidence for idealism, not the *physical* one.

Ultimately, given the experimental confirmation of quantum mechanical contextuality, the articulation in Chapter 6—though less intuitive—should be closer to the truth than that in Chapter 5.

Chapter 5

An ontological solution to the mind-body problem

This article first appeared in *Philosophies*, ISSN: 2409-9287, Vol. 2, No. 2, Article No. 10, on 20 April 2017. *Philosophies* is published by MDPI AG, Basel, Switzerland. According to the website openaccess.nl, sponsored by Dutch universities and research institutes to foment the publication of publicly-funded research in open-access journals, MDPI AG was one of the most popular open-access publishers amongst Dutch academics in 2016.[1]

5.1 Abstract

I argue for an idealist ontology consistent with empirical observations, which seeks to explain the facts of nature more parsimoniously than physicalism and bottom-up panpsychism. This ontology also attempts to offer more explanatory power than both physicalism and bottom-up panpsychism, in that it does not fall prey to either the 'hard problem of consciousness' or the 'subject combination problem,' respectively. It can be summarized as follows: spatially unbound consciousness is posited to be nature's sole ontological primitive. We, as well as all other living organisms, are dissociated alters of this unbound consciousness. The universe we see around us is the extrinsic appearance of phenomenality surrounding—but dissociated from—our alter. The living organisms we share the world with are the extrinsic appearances of other dissociated alters. As such,

---

1 See: http://openaccess.nl/en/what-is-open-access/open-access-publishers (accessed on 24 April 2017). The Netherlands is my home country and the location of my *alma mater*.

the challenge to artificially create individualized consciousness becomes synonymous with the challenge to artificially induce abiogenesis.

5.2 Introduction

The mind-body problem—that is, the question of how conscious experience relates to arrangements of matter—is inextricably tied to ontology. The mainstream physicalist ontology, for instance, posits that reality is constituted by irreducible entities—which I shall call 'ontological primitives,' or simply 'primitives'—outside and independent of experience. According to physicalism, these primitives, in and of themselves, do not experience. In other words, there is nothing it is like to *be* a primitive, experience somehow emerging only at the level of complex arrangements of primitives. As such, under physicalism experience is not fundamental, but instead reducible to physical parameters of arrangements of primitives. What I shall call 'microexperientialism,' in turn, posits that there is already something it is like to be at least some primitives (Strawson et al. 2006: 24-29), combinations of these experiencing primitives somehow leading to *more complex* experiences. As such, under microexperientialism experience is seen as an irreducible aspect of at least some primitives. The ontology of panexperientialism (Griffin 1998: 77-116, Rosenberg 2004: 91-103, Skrbina 2007: 21-22) is analogous to microexperientialism, except in that the former entails the stronger claim that *all* primitives experience. Finally, micropsychism (Strawson et al. 2006: 24-29) and panpsychism (Skrbina 2007: 15-22) are analogous—and, in fact, may be identical—to microexperientialism and panexperientialism, respectively, except perhaps in that some formulations of the former admit cognition—a more complex form of experience— already at the level of primitives, as an irreducible aspect of these primitives. For ease of reference, I shall henceforth group microexperientialism, panexperientialism, micropsychism and panpsychism, as defined above, under the label 'bottom-up panpsychism.'

If we stipulate that an entity is *conscious* if, and only if, there is something — *anything* — it is like to *be* the entity, we can then summarize the discussion above as follows: (a) physicalism posits that all ontological primitives, in and of themselves, are *un*conscious, consciousness arising only at the level of complex arrangements of primitives; (b) bottom-up panpsychism posits that at least some ontological primitives are *conscious* in and of themselves, their combinations leading to more complex consciousness.

Notice, however, that the question of what physical entities are or are not conscious is not the only angle through which to approach the mind-body problem. Indeed, according to the ontology of idealism, physical entities exist only insofar as they are *in consciousness*, irrespective of whether they are conscious or unconscious. In other words, whilst physicalism and bottom-up panpsychism entail that there are physical entities or arrangements thereof that *circumscribe* consciousness, idealism posits that all physical entities and arrangements thereof are *circumscribed by* consciousness. This is a significant distinction that alone sets idealism — whatever its particular formulation — apart from all other ontologies discussed.

The present paper seeks to derive the simplest and most explanatorily powerful ontology possible from the basic facts of reality, thereby attempting to solve the mind-body problem. It starts by stating these basic facts precisely, in a way that avoids any *a priori* metaphysical assumption or bias. A series of inferences are then made, based on empirical rigor, logical consistency and parsimony. These inferences ultimately lead to an idealist ontology that explains all the basic facts. Explicit comparisons are finally made between the ontology so derived and those of physicalism and bottom-up panpsychism, in terms of both parsimony and explanatory power.

Before we begin, however, notice that idealism has a long and rich history, which can be traced back to the Vedas in the East and Neoplatonism in the West. Many different schools of idealism are known today, such as 'subjective idealism,' 'absolute idealism,' 'actual idealism,' etc. The criteria for classifying a new formulation under one or another school are often difficult to apply with precision, due to their often ambiguous definitions and inconsistent usage of words such as 'mind,' 'consciousness,' 'experience,' 'subject,' 'object,' etc. For this reason, I have chosen to simply present my approach in and of itself. Others can worry later about classifying it, if they find it worthwhile.

5.3 The basic facts of reality

Let us start by neutrally and precisely stating four basic facts of reality, verifiable through observation, and therefore known to be valid irrespective of theory or metaphysics:

Fact 1: There are tight correlations between a person's reported private experiences and the observed brain activity of the person.

We know this from the study of the neural correlates of consciousness (e.g. Koch 2004).

Fact 2: We all seem to inhabit the same universe.

After all, what other people report about their perceptions of the universe is normally consistent with our own perceptions of it.

Fact 3: Reality normally unfolds according to patterns and regularities—that is, the laws of nature—independent of personal volition.

Fact 4: Macroscopic physical entities can be broken down

into microscopic constituent segments, such as subatomic particles.

What makes these four particular facts significant is this: despite the formidable unresolved problems of both physicalism (Levine 1983, Chalmers 2003, Nagel 2012, Kastrup 2014, Kastrup 2015) and bottom-up panpsychism (Goff 2009, Coleman 2014, Chalmers 2016), these two ontologies are *prima facie* more easily reconcilable with the four facts than idealism.

On the physicalist side, the argument for this might go as follows: If the brain doesn't somehow constitute or generate conscious experience through specific arrangements of its microscopic constituent segments (Fact 4), how can there be such tight correlations between observed brain activity and reported inner experiences (Fact 1)? If the world isn't fundamentally independent of, and outside, phenomenality, it can only be analogous to a dream in consciousness. But in such a case, how can we all be having the same 'dream' (Fact 2)? Finally, if the world is in consciousness, how can it unfold according to patterns and regularities independent of our personal volition (Fact 3)?

On the bottom-up panpsychist side, the following considerations might be added to the above: Since physicalism has hitherto failed to explain how the qualities of experience can be deduced from physical parameters, experience must be fundamental. The question then is: fundamental at what level? Well, since the macroscopic brain can be reduced to microscopic building blocks (Fact 4), experience must be a fundamental aspect of these microscopic building blocks.

5.4 Unpacking the basic facts

By carefully unpacking Fact 1, we can confidently state five other facts:

Fact 5: Irrespective of the ontological status of what we call 'a person,' there is *that* which experiences (TWE).

Properly understood, this is self-evident and, as colorfully put by Strawson (2006: 26), not even a sensible Buddhist rejects such a claim. For clarity, notice that I am not necessarily making an ontological distinction between experience and experienc*er* here; in fact, soon I will claim precisely that there isn't such a distinction. I am simply recognizing that experience necessarily entails a subjective field of potential or actualized qualities. TWE *is* this field.

Notice also that I am not, at least for now, passing any judgment or making any assumption about the nature or boundaries of TWE. I am not saying, for instance, that it is or isn't physical, or spiritual, or informational, etc. I am not saying that it is or isn't circumscribed by the skin of a higher animal. I am simply asserting that it inevitably exists, whatever its nature may be and wherever its boundaries may lie.

Fact 6: A person has private experiences that can only be known by others if the person reports them, for other people do not have direct access to these private experiences.

Fact 7: The brain activity of a person is known only insofar as its observation is experienced in the form of perceptions.

For instance, if a neurologist performs a functional magnetic resonance imaging (fMRI) scan or an electroencephalogram (EEG) of a person's brain activity, the measurements are only known insofar as the neurologist—or someone else—*sees* them consciously.

Fact 8: From Facts 1 and 7, there are tight correlations between two types of experience: (a) conscious perceptions of a person's brain activity and (b) private experiences of the person.

Let us call these the *extrinsic appearance* and the *intrinsic view*, respectively. More generally, the intrinsic view is an entity's conscious inner life, while the extrinsic appearance is how this conscious inner life is perceived by another entity e.g. through instrumentation. Both the intrinsic view and the extrinsic appearance are, of course, still *experiences* insofar as they can be known.

Fact 9: A brain has the same essential nature — that is, it belongs to the same ontological class — as the rest of the universe.

After all, brains are made of the same kind of 'stuff' that makes up the universe as a whole.

5.5 Deriving an idealist ontology from the basic facts

The question that presents itself now is this: What is the most parsimonious ontological explanation for these nine facts? Here I use the qualifier 'parsimonious' in the sense of Occam's Razor: the most parsimonious ontology is that which requires the smallest number of postulates whilst maintaining sufficient explanatory power to account for all facts. In what follows, I offer six inferences that, together, aim to answer this question.

Inference 1: The most parsimonious and least problematic ontological underpinning for Fact 5 is that TWE and experience are of the same essential nature. More specifically, experience is a *pattern of excitation* of TWE.

This avoids the need to postulate two different ontological classes for TWE and experience, respectively. It also circumvents problems regarding the mechanisms of interaction between TWE and experience, which would arise if they were assumed to be of different essential natures. As an excitation of TWE, experience is not distinct from TWE as ripples are not distinct from water, or as a dance is not distinct from the dancer. There is nothing to ripples but water in motion. There is nothing to a dance but the dancer in motion. In an analogous way, there is nothing to experience but TWE 'in motion.' Ripples, dances and experience are merely patterns of excitation of water, dancers and TWE, respectively.

Now, from Fact 8 we know that the activity of brains is accompanied by inner experience. In other words, there is something it is like to *be* a living brain. One possibility is that something about the particular structure or function of brains constitutes or generates experience. However, it is impossible to conceive—even *in principle*—of how or why any particular structural or functional arrangement of physical elements would constitute or generate experience (Rosenberg 2004: 13-30, Strawson et al. 2006: 2-30). This is a well-known problem in neuroscience and philosophy of mind, often referred to as the 'hard problem of consciousness.' The qualities of experience are irreducible to the observable parameters of physical arrangements—whatever the arrangement is—in the sense that it is impossible to deduce those qualities—even in principle—from these parameters (Chalmers 2003). It remains conceivable that physical arrangements could *modulate* experience, without constituting or generating it, if one postulates some form of dualism. But this still leaves 'that which experiences' entirely unexplained, since TWE is now that which is modulated (cf. Inference 1). From all this we can conclude that:

Inference 2: TWE is an ontological primitive, uncaused and irreducible.

Clearly, this step of my argument depends on the 'hard

problem' being a fatal blow to the notion that physical stuff more fundamental than experience somehow constitutes or generates experience. There is now, of course, substantial literature supporting this view (e.g. Levine 1983, Chalmers 2003, Rosenberg 2004: 13-30, Strawson et al. 2006: 2-30, etc.). Nonetheless, you may still disagree with Inference 2 for two reasons: (a) you may think that physicalism in fact does not entail a 'hard problem' (e.g. Dennett 2003); or (b) you may think that the 'hard problem' *can* be solved, even though today we do not know how. Position (a) implies that conscious experience essentially does not exist, which, as I have extensively argued elsewhere (Kastrup 2015: 59-70), is absurd. After all, conscious experience—whatever its underlying nature—is the primary datum of existence. Position (b), on the other hand, cannot be refuted upfront because, outside closed formal systems such as mathematics or logic, one often cannot prove a negative. But if you sympathize with position (b), my invitation to you is this: continue nonetheless to entertain my argument to its conclusion; compare physicalism to the idealist ontology that will emerge from it at the end; and then ask yourself which alternative is more parsimonious.

Having briefly digressed, let us now proceed. Since 'that which experiences' cannot be caused by local physical arrangements (Inference 2), and since living brains—which *do* experience (Fact 8)—are of the same essential nature as the rest of the universe (Fact 9), we must face the possibility that the latter also experiences. Rejecting this conclusion entails accepting an arbitrary discontinuity in nature. As such, the entire physical universe may be akin to a 'nervous system' in the specific sense that all its activity may be accompanied by experience. Is there any circumstantial empirical evidence for this kinship? As it turns out, there is: a study has shown unexplained structural similarities—not necessarily *functional* ones, mind you—between the universe at its largest scales and biological brains (Krioukov et al. 2012).[2] We can thus cautiously attempt:

Inference 3: TWE is associated with the entire universe.

This does not imply that the activity of particular subsets of the universe is accompanied by *separate* conscious inner lives of their own. Asserting otherwise would require an extra inferential step. As such, it cannot be logically concluded from Inference 3 that there is something it is like to be, say, a home thermostat in and of itself. To gain intuition about this, consider e.g. an individual neuron in your brain: Is there anything it is like to be it, in and of itself? Insofar as you can directly experience, there isn't: there is only something it is like to be your brain *as a whole*—that is, you—not the individual neuron in and of itself. Nonetheless, this observation does not contradict Fact 8: the activity of the neuron is still accompanied by experience, *but experience at the level of your brain as a whole*. Analogously, Inference 3 must be interpreted parsimoniously as implying solely that all activity in the physical universe is accompanied by conscious inner life *at some level*, and not necessarily that particular subsystems of the universe—such as home thermostats—have separate conscious inner lives at their own level.

The best that can be concluded beyond this cautious interpretation of Inference 3 is that TWE is, in fact, *unitary* at a universal level: the validity of the laws of nature across time and space seem to indicate a holistic underlying reality, as opposed to a fundamentally fragmented one. Moreover, as argued by Schaffer, "there is good evidence that the cosmos forms an entangled system, and good reason to treat entangled systems as irreducible wholes" (2010: 32). Horgan and Potrč had already arrived at similar conclusions earlier (2000). So if the cosmos is

---

2 This conclusion has been confirmed and amplified by a later study done by Franco Vazza and Alberto Feletti (2017).

an irreducible whole, then TWE—which is associated with the entire cosmos, as per Inference 3—must be unitary.

Yet, we know empirically that living people have separate, private experiences (Fact 6). Many of my personal experiences are surely not the same as yours. Moreover, I am not aware of what is going on in the universe as a whole and, presumably, neither are you. To reconcile these facts with the discussion above, I propose as a useful analogy a common mental condition called *dissociation*. Dissociative states are well recognized in psychiatry today, featuring prominently in the DSM-5 (American Psychiatric Association 2013). Their hallmark is "a disruption of and/or discontinuity in the normal integration of consciousness, memory, identity, emotion, perception" (Black & Grant 2014: 191). In other words, dissociation entails that some mental contents cannot evoke other mental contents, leading to *apparent* fragmentation. A person suffering from a particularly severe form of dissociation called Dissociative Identity Disorder exhibits multiple, "discrete centers of self-awareness" (Braude 1995: 67) called *alters*.

Dissociation allows us to (a) grant that TWE is fundamentally unitary at a universal level and then still (b) coherently explain the private character of our personal experiences (Fact 6):

Inference 4: There is a sense in which living organisms are alters of unitary TWE.

It is important to notice that the formation of alters does not entail or imply fragmentation of TWE itself, but only the dissolution of cognitive bridges between some of TWE's mental contents. Even when these mental contents are dissociated from each other—in the sense of not being able to directly evoke each other—TWE remains unitary. Let us unpack this.

As mentioned above, dissociation entails "a disruption of and/or discontinuity in the normal integration" of mental contents. This normal integration takes place through chains of cognitive associations: a perception may evoke an abstract idea, which may trigger a memory, which may inspire a thought, etc. These associations are *logical*, in the sense that e.g. the memory inspires the thought because of a certain *implicit logic* linking the two. Integrated mentation can thus be modeled, for ease of visualization, as a connected, directed graph. See Figure 5.1a. Each vertex in the graph represents a particular mental content and each edge a cognitive association logically linking mental contents together.[3] Every mental content in the graph of Figure 5.1a can be reached from any other mental content through a chain of cognitive associations. Dissociation, in turn, can be visualized as what happens when the graph becomes disconnected, such as shown in Figure 5.1b. Some mental contents can then no longer be reached from others. The inner subgraph is thus a representation of an alter.

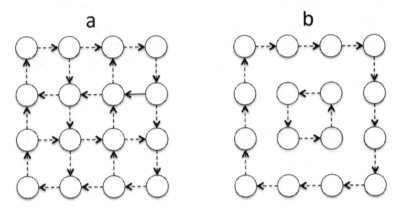

Figure 5.1: A connected graph (a) illustrating normal integration of mental contents, and a disconnected graph (b) illustrating dissociation and the corresponding formation of an alter (inner subgraph).

An alter loses access to—that is, the power to evoke—mental contents surrounding it, *but remains integral to TWE*. The disconnection between an alter and the surrounding mental contents is logical, not ontic. As an analogy, a database may contain entries that are not indexed and, therefore, cannot be reached, but this does not physically separate those entries from the rest of the database. Similarly, dissociation allows us to explain the existence of separate, private conscious inner lives, whilst preserving the notion that TWE is, and always remains, fundamentally unitary.[4]

As discussed above, the empirical motivation for positing dissociation as the explanation for Fact 6 is the clinical condition called Dissociative Identity Disorder (DID). There has been

---

3 Since a mental content is an experience, each vertex in the graphs of Figure 5.1 represents a particular pattern of excitation of TWE. Each edge in the graphs represents thus an association between two patterns of excitation, each pattern with its particular constituent harmonics. When the two patterns of excitation are concurrently present—that is, when the two associated mental contents are experienced together—the association can be seen as a combination of the respective harmonics, like in a musical chord wherein multiple notes are played at the same time. When the association unfolds in temporal sequence—as e.g. in the case of a thought that fades away to make room for the experience of the memory it evokes—it can be visualized as a transition from the first to the second pattern of excitation, like notes played in sequence in a melody.

4 In a discussion I had with Daniel Stoljar in China, in June of 2017, Daniel pointed out that alters cannot concurrently *be* TWE *and* have distinct experiences. After all, if alters have different experiences they must be distinct from one another; but if all alters are ultimately TWE itself, then they cannot be distinct. The key to resolving this apparent contradiction is to understand that, although there is nothing to an alter but TWE itself, there is more to TWE than any particular one of its alters. Daniel's suggestion, which I embrace, is thus to think of alters as 'parts' of TWE—local differentiations of their common substrate—but in a way that doesn't imply a commitment to TWE having spatial or temporal extension. This way, alters can be regarded as parts of TWE in the generic and non-committal sense that e.g. a mathematical equation can have different parts despite having no spatial or temporal extension.

debate about the authenticity of DID. After all, it is conceivable that patients could fake it. Research, however, has confirmed DID's legitimacy (for an overview, see Kelly et al. 2009: 167-174). Two very recent studies are particularly interesting to highlight. In 2015, doctors reported on the case of a German woman who exhibited a variety of alters (Strasburger & Waldvogel). Peculiarly, some of her alters claimed to be blind while others could see normally. Through EEGs, the doctors were able to ascertain that the brain activity normally associated with sight wasn't present while a blind alter was in control of the woman's body, even though her eyes were open. When a sighted alter assumed control, the usual brain activity returned. This is a sobering result that shows the literally *blinding* power of dissociation.

In another study (Schlumpf et al. 2014), doctors performed fMRI brain scans on both DID patients and actors simulating DID. The scans of the actual patients displayed clear and significant differences when compared to those of the actors. This study is interesting not only for confirming the authenticity of DID, but also for showing that *dissociation has an extrinsic appearance*. In other words, there is something dissociative processes *look like* when observed from the outside, through a brain scanner. The significance of this fact will become clear shortly.

Finally, there is also compelling evidence that alters can remain conscious and self-aware even when not in control of the body. In Morton Prince's well-known study of the 'Miss Beauchamp' case of DID, one of the alters "was a co-conscious personality in a deeper sense. When she was not interacting with the world, she did not become dormant, but persisted and was active" (Kelly et al. 2009: 318). Braude's more recent work corroborates the view that alters can be co-conscious. He points to the struggle of different alters for executive control of the body and the fact that alters "might intervene in the lives of others [i.e., other alters], intentionally interfering with their interests and activities, or at least playing mischief on them" (1995: 68). It thus appears that alters can not only be concurrently conscious, but that they can also vie for dominance with each other.

As seen above, dissociation is an empirically established phenomenon known to occur in experiential space, which can lead to the formation of co-conscious alters. And since TWE is universal experiential space (Inference 3), it is empirically coherent to posit—as Inference 4 does—that top-down dissociation leads to the formation of discrete but concurrently conscious centers of experience within the otherwise unitary TWE.

The challenge we must now tackle is the so-called "boundary problem for experiencing subjects" (Rosenberg 2004: 77-90): What structures in nature correspond to alters of TWE? We know that we humans do. Do animals too? What about plants? Rocks? Atoms? Subatomic particles?

As Gregg Rosenberg put it, "we must find something in nature to ground [the boundaries of] an experiencing subject" (2004: 80)—that is, the outline of the extrinsic appearance of an alter of TWE on the screen of perception. This "something in nature" must have structural and functional characteristics that allow us to differentiate it from everything else. After all, only on the basis of this differentiation can we delineate the dissociated alters from an extrinsic perspective. But just what is the structure Rosenberg was looking for? Departing here from Rosenberg's own conclusions, I posit that a natural and empirically plausible candidate is metabolizing life:

Inference 5: Metabolizing organisms are the extrinsic appearance of alters of TWE.

The reasoning here is simple: since we only have intrinsic access to ourselves, we are the only structures *known* to have dissociated streams of inner experiences. We also have good empirical reasons to conclude that normal metabolism is essential for the maintenance of this dissociation, for when it slows down or stops the dissociation seems to reduce or end (Kastrup 2017a[5]). These observations alone suggest strongly that metabolizing life is the structure corresponding to alters of TWE.

But there is more: insofar as it resembles our own, the extrinsic behavior of *all* metabolizing organisms is also suggestive of their having dissociated streams of inner experiences analogous to ours in some sense. This is obvious enough for cats and dogs, but—you might ask—what about plants and single-celled organisms such as amoebae? Well, consider this: "many types of amoeba construct glassy shells by picking up sand grains from the mud in which they live. The typical *Difflugia* shell, for example, is shaped like a vase, and has a remarkable symmetry" (Ford 2010: 26). As for plants, many recent studies have reported their surprisingly sophisticated behavior, leading even to a proposal for a new field of scientific inquiry boldly called "plant neurobiology" (Brenner et al. 2006). Clearly, thus, even plants and single-celled organisms exhibit extrinsic behavior somewhat analogous to our own, further suggesting that they, too, may have dissociated streams of inner experiences. Of course, the same cannot be said of any inanimate object or phenomenon (those that have been engineered by humans to merely simulate the behavior of living beings, such as robots, natural language interfaces, etc., naturally don't count).

Finally, there is no doubt that metabolism is a highly differentiated process. Consider DNA, morphogenesis, transcription, protein folding, mitosis, etc.: nothing else in nature exhibits structural and functional characteristics such as these. And it is these characteristics that unify all metabolizing life into a unique, clearly distinct natural category, despite the widely different forms that organisms can take. This category may provide the unambiguously demarcated "something in nature" that Rosenberg was looking for.

---

5 This reference can be found in Chapter 11 of the present volume.

The essence of Inference 5 is that there is something an alter of TWE looks like from outside; namely, a metabolizing body. By now this shouldn't come as a surprise: recall that, in the discussion leading to Inference 3, I've posited that the physical universe is, in a specific sense, akin to a 'nervous system.' Recall also that a study has shown that dissociative processes in the nervous systems of DID patients have a distinct extrinsic appearance, detectable by brain scans (Schlumpf et al. 2014). Therefore, it is plausible that dissociation in the universal 'nervous system' should also have a distinct extrinsic appearance. The hypothesis here is that metabolizing organisms *are* this extrinsic appearance. As such, living bodies are to universal-level dissociation in TWE as certain patterns of brain activity are to DID patients. In the case of the universal 'nervous system,' however, we don't need brain scanners, for we are already *inside* the 'nervous system.' To see the extrinsic appearance of dissociated mental processes within it we just need to look around: the people, cats, dogs, insects, plants, amoebae and all other life forms we see around are the diagnostic images of universal 'DID.' Each corresponds to at least one alter.

For clarity and emphasis, notice that I have been elaborating on two levels: TWE as a whole and its dissociated alters, which are themselves nothing but local differentiations of TWE. Moreover, there are two ways in which an alter of TWE can be experienced: (a) its *extrinsic appearance*—that is, the metabolizing organisms we can perceive around us; and (b) its *intrinsic view*, an example of which is your own stream of inner experiences as an alter yourself. Moreover, unless we are prepared to accept an arbitrary discontinuity in nature, the same must apply to the rest of the universe: its extrinsic appearance is the cosmos we perceive around us, while its intrinsic view is the hypothetical stream of inner experiences of TWE as a whole.

One may feel tempted to conclude that this implies some form of dual-aspect monism, *a la* Spinoza (Skrbina 2007:

88), whereby intrinsic views and extrinsic appearances are irreducible to one another. What I shall attempt to show next is that this is not so: extrinsic appearances can in fact be reduced to intrinsic views.

Before I continue, however, notice that it is *perceptions* that carry extrinsic appearances, not thoughts (for simplicity, I shall henceforth use the word 'thought' to refer to any experience distinct from perception). If all you experienced were thoughts, you would have no extrinsic point of view at all, only an intrinsic one. Therefore, if I can coherently reduce perceptions at the level of alters to thoughts at the level of TWE as a whole, I will have shown that nature, at its most fundamental level, consists purely of intrinsic views.

With this in mind, I submit that, before its first alter ever formed, TWE experienced *only thoughts*. There were no perceptions. The formation of the first alter then demarcated a boundary separating the experiences *within* the alter from those *outside* the alter (all of which were, of course, still within TWE). This newly formed boundary is what enabled perceptions to arise: the thoughts surrounding the alter *stimulated* its boundary from the outside, which in turn impinged on the alter's internal dynamics. What we call perception is the experience of this impingement (see Figure 5.2).[6] Naturally, the thoughts of the alter can also stimulate its dissociative boundary from the inside and thereby impinge on the external dynamics of TWE (not shown in Figure 5.2). This corresponds to the effects on the world of the presence and actions of a living organism within it, which cause something akin to perception in the external mental

---

6 Insofar as experiences are excitations of TWE, this impingement can be visualized as an *interference pattern* between excitations originating within and outside the dissociative boundary of the alter, respectively. Perceptual experiences then correspond to these interference patterns. This notion is elaborated on further in Chapter 6 of this volume.

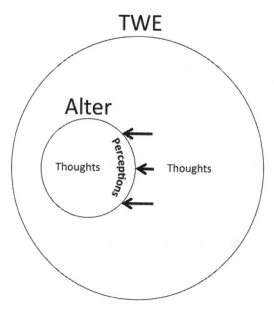

Figure 5.2: Thoughts in TWE cause perceptions in a dissociated alter.

environment surrounding the alter. For the sake of simplicity, however, I shall henceforth talk of perception only as it applies to alters.

Restating the key point more generally:

> *Inference 6*: The perceptions of an alter are reducible to the experiences of TWE that impinge on the alter from the outside.

The extrinsic appearance of an alter's boundary is, of course, an organism's sense organs. In our case, these are our skin, eyes, ears, nose and tongue. Therefore, even if the outside stimulation is very faint and subtle, evolution has had billions of years to optimize the sensitivity of our sense organs—our alters' boundaries—to pick up on these hypothetically faint signals.

Two questions can be raised at this point: First, how can a mere dissociative boundary give rise to a *qualitatively different* type of experience? After all, perceptions *feel* very distinct from thoughts. Second, how can outside experiences, which are by definition dissociated from the alter, cause experiences inside the alter? This seems contradictory at first.

Let us start from the second question. Contrary to the question's premise, we are all, in fact, personally familiar with dissociated experiences that causally affect each other while remaining dissociated from each other. Imagine, for instance, that you are having relationship problems at home. When you go to work, you successfully 'park' your problems—that is, repress your emotional life—in order to perform your tasks. Your emotions then become temporarily dissociated from your ego, in the sense that they are no longer evoked in your awareness while you work. *But they do still impinge on it*: they may, for instance, cause your imagination to flow in a somber direction, lead you to misunderstand comments received from colleagues, lock your intellect into repetitive patterns of reasoning, etc.[7] All the while, your ego doesn't directly experience the emotions themselves; they remain dissociated from it. But from across the dissociative

---

7 This is an important point. If two mental contents A and B—that is, particular experiences or patterns of excitation of TWE—are dissociated from one another, this means only that they cannot evoke one another. But B can, in principle, still *influence* A in the sense that the presence of B somewhere in TWE may qualitatively interfere with how A feels, even when A cannot evoke B in its own segment of TWE. Indeed, in the context of a vibration metaphor, B is dissociated from A when the pattern of excitation corresponding to A in a given segment of TWE neither comprises nor is followed by the constituent harmonics of B in that segment of TWE. Nonetheless, if B unfolds in parallel with A in a *different* segment of TWE, its corresponding pattern of excitation can, in principle, still *interfere* with that of A at a shared dissociative boundary, like two chains of ripples in water interfere with one another at the point where they meet.

boundary they still causally influence what arises in your egoic awareness. Indeed, the plausibility of this kind of phenomenal impingement from across a dissociative boundary is well established in the literature: Lynch and Kilmartin (2013: 100), for instance, report that dissociated feelings can dramatically affect our thoughts and behaviors, while Eagleman (2011: 20–54) shows that dissociated expectations routinely mold our perceptions. My claim is that something analogous to this happens across the boundary of dissociated alters of TWE, causing perceptions.

Let us now tackle the first question. Still with reference to the example described above, notice that your dissociated emotions at work have an impact on *qualitatively different* types of experience: they interfere with your imagination, understanding and reasoning, none of which *feels* like emotions. This shows empirically that, not only can there be a causal link across a dissociative boundary, this causal link can also connect qualitatively dissimilar experiences. A dissociated emotion can influence a thought; a dissociated belief can distort a perception through hallucinatory mechanisms; etc. Therefore, it is empirically coherent to infer that experiences outside an alter can causally influence qualitatively dissimilar experiences in the alter. It is reasonable to postulate even that evolution would have *emphasized* this kind of qualitative transition, if it helped enhance the sensitivity of the alter to external stimuli.[8]

As a matter of fact, these empirically-motivated speculations can be couched in recent theoretical results. Donald Hoffman's interface theory of perception (2009, Hoffman & Singh 2012), for

---

8 In the context of a vibration metaphor, we have seen that a perception is an interference pattern between thoughts respectively inside and outside an alter. And we know that an interference pattern may differ significantly from the original patterns of excitation that give rise to it. This difference may help us visualize why perceptions are qualitatively so unlike thoughts.

instance, shows that evolution emphasizes perceptual qualities conducive to fitness, not to truth. In other words, we have evolved to perceive not the qualities that are really 'out there' — that is, outside our alter — but just a *representation* thereof that helps us to survive and reproduce. Hoffman uses the analogy of a computer desktop: although a computer file is represented in it as e.g. a blue rectangle, this does not mean that the file itself has the qualities of being blue and rectangular. As a matter of fact, the actual file does not have those qualities at all: it is a pattern of open and closed microscopic switches in a silicon chip. In an analogous way, my hypothesis is that the qualities of our perceptions — colors, shapes, sounds, flavors, textures, etc. — are *not* the qualities of the experiences of TWE that surround our alter, but their 'desktop representation' instead. Our perceptions do not feel like the thoughts of TWE (see Figure 5.2 again) because a *qualitative transition* between these two experiential categories has helped our ancestors survive and reproduce.

The work of Friston, Sengupta and Auletta (2014) has similar implications but, significantly, is derived from an entirely different line of reasoning. Their results are based on abstract mathematical considerations and, therefore, can in principle be leveraged under any ontology. They show that whenever a boundary — a "Markov Blanket" in their mathematical model — defines the outline of an individual organism, internal states of the organism will evolve to optimize for two conflicting goals: (a) to reflect external states of the world beyond the Markov Blanket; and (b) to minimize their own entropy or dispersion. Goal (a) is about allowing the organism to know what is going on in the world outside, so it can take suitable actions to survive in that world. Goal (b) is about preventing the organism from losing its internal structural and dynamical integrity because of the second law of thermodynamics. Naturally, in our case the Markov Blanket is the dissociative boundary of an alter, whose extrinsic appearance is our skin and other sense organs.

Dissociation thus creates a Markov Blanket within TWE.

The key insight of Friston, Sengupta and Auletta can be paraphrased as follows: a hypothetical organism with perfect perception—that is, able to perfectly *mirror* the qualities of the surrounding external world in its internal states—would not have an upper bound on its own internal entropy, which would then increase indefinitely. Such an organism would dissolve into an entropic soup. To survive, organisms must, instead, use their internal states to actively *represent* relevant states of the outside world in a *compressed, coded form*, so to know as much as possible about their environment while remaining within entropic constraints compatible with maintaining their structural and dynamical integrity. This way, my hypothesis is that the qualities of perception experienced by an alter are just compressed, coded representations of how surrounding thoughts of TWE actually feel. As such, while there must be a *correspondence* between perception and surrounding thoughts, the respective experiential qualities don't need to be the same. In fact, they will likely be very different if it helps organisms resist entropy. Our perceptions don't feel like thoughts because they are coded representations thereof.

These six inferences complete the proposed idealist ontology. We must now check how well this ontology explains the four basic facts of reality that we started with in Section 5.3.

5.6 Explaining the basic facts of reality

I showed in Section 5.3 how Facts 1 to 4 can be construed *prima facie* to favor physicalism and bottom-up panpsychism over idealism, despite the formidable unresolved problems of the former. What I hope to show next is that, in fact, the idealist ontology articulated in Inferences 1 to 6 can explain those four facts at least as elegantly. Moreover, as I shall discuss later, the proposed ontology does not suffer from the problems that physicalism and bottom-up panpsychism fall prey to.

In what follows, each explanation is numbered according to the fact it explains. For instance, Explanation 1 explains Fact 1.

Explanation 1: Let us start by noticing that, from an empirical perspective, there is nothing to Fact 1 that is not captured in Fact 8. Therefore, by explaining Fact 8 we also explain Fact 1. From Inference 6, for any given alter *A1* of TWE, it is the experiences surrounding *A1* that cause its perceptions of the world around it. Naturally, dissociated experiences corresponding to another alter *A2* can be part of the experiential environment surrounding *A1*. As such, the inner experiences of *A2* can also indirectly stimulate *A1*'s boundary—by impinging on their shared experiential environment—and thereby cause *A1*'s perceptions of *A2*. This is what gives *A1* an *extrinsic* view of the inner experiences of *A2* in the form of *A2*'s metabolizing body (Inference 5). See Figure 5.3. And since *A2*'s brain is an integral part of its body, it follows that *A2*'s inner experiences *cause* the perception by *A1* of the activity in *A2*'s brain. This causal link explains Fact 8 and, therefore, Fact 1.

Putting it more generally, the extrinsic appearance and intrinsic view of an organism correlate tightly with one another because the intrinsic view causes the extrinsic appearance, not the other way around. Contrary to physicalism, thus, it is the inner experiences of an organism—including non-self-reflective and internally dissociated types unreachable through introspection, which I shall elaborate upon in Section 5.7[9]—that cause its body (see also Kastrup 2015: 17-18 & 189-190), insofar as the body is no more than a set of perceptions. (Notice that *A1* and *A2* can also be the *same* alter, since an organism can perceive its own body.)

---

9 Chapter 9 elaborates on these kinds of experience much more extensively.

TWE

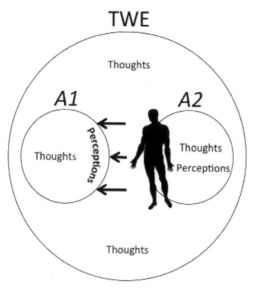

Figure 5.3: The dissociated experiential activity of an alter *A2* can also stimulate the boundary of another alter *A1* by impinging on their shared experiential milieu.

Explanation 2: Since TWE is universal (Inferences 3 and 4), it follows that all alters of TWE—that is, metabolizing organisms such as ourselves (Inferences 4 and 5)—are immersed, like islands of a single ocean, in the thoughts that constitute the intrinsic view of the non-metabolizing segment of the universe. These universal thoughts surround all alters and cause their perceptions by stimulating their respective dissociative boundaries (Inference 6). See Figure 5.4.

Moreover, since the thoughts of TWE are excitations of TWE itself (Inference 1), it follows trivially that we can explain our shared universe based on excitations of TWE alone.

Explanation 2 may raise a plausibility objection, since it entails that thoughts in TWE as a whole need to be significantly more orderly than those in our personal psyches. I deal with this objection in the next section. For now, let us continue.

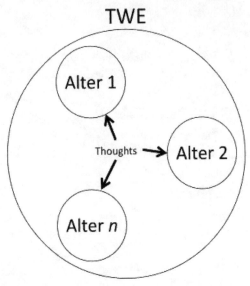

Figure 5.4: Our shared universe arises from the thoughts of TWE in which all alters are immersed.

Explanation 3: Since volition is innately experiential, the volition of each and every alter of TWE is also dissociated from the rest of TWE (Inference 4). This explains rather simply why we do not have personal volitional control over the laws of nature: the unfolding of the universe around ourselves consists of excitations of TWE from which we are dissociated.

Explanation 4: The perceptions of an alter are coded representations of experiences in TWE that surround the alter (Inference 6). Those experiences in TWE are excitations or 'movements' of TWE itself (Inference 1). Therefore, subatomic particles, as the smallest discernible elements or 'pixels' of the perceived world, are coded representations of the smallest discernible 'movements' of TWE.

Facts 1 to 4—and, in fact, Facts 5 to 9 as well—are now explained in terms of TWE. Naturally, from Inference 2 we know that we don't need to explain TWE itself: it is an ontological primitive. Indeed, every theory of nature needs to identify at least one ontological primitive, since we cannot keep on explaining one thing in terms of another forever. At some point we have to stop and say: at this level, nature simply is. So TWE *simply is*. The fact that I do not reduce it to something else is in no way a shortcoming. Physicalism and bottom-up panpsychism themselves, depending on their specific formulation, postulate a slew of abstract subatomic particles, the quantum field, 'superstrings,' 'branes,' etc. as primitives, all of which are also fundamentally unexplainable. If anything, my formulation has the advantages of both parsimony—for making do with a single primitive—and empirical rigor—for choosing an indisputable empirical fact as primitive, as opposed to abstractions of thought.

The proposed ontology thus reduces everything to TWE, its sole ontological primitive. And as attentive readers have probably already noticed, 'consciousness' is the ordinary English word that best fits what is meant by TWE:

TWE = universal consciousness

5.7 Criticisms of the proposed ontology

In another work (Kastrup 2017c[10]), I have rebutted several objections to idealism. It is not in the scope of this essay to repeat all those rebuttals, but three particularly relevant objections must be anticipated and addressed here.

The first objection relates to Libet's experiments (1985), wherein neuroscientists were able to record, a fraction of a second *before* subjects reported making a decision to act, mounting brain activity associated with the initiation of a simple voluntary action. At first sight, this would seem to indicate that decisions are made in a neural substrate outside consciousness, thereby

---

10 This reference can be found in Chapter 8 of the present volume.

contradicting the proposed ontology.

The fallacy behind this objection is a conflation of consciousness itself with a particular *configuration of* consciousness. But before I get to it, notice that the proposed ontology entails that dissociation can happen in nested levels: TWE can dissociate into multiple people, and then a person suffering from DID can, in turn, dissociate into multiple personal alters. Dissociation within a person does not even require pathology, for there is significant evidence that we all have a second center of cognition—an 'other'—dissociated from our ego (Kelly et al. 2009: 301-365). Indeed, this is the very foundation of depth psychology and relatively recent results in neuroscience corroborate it (Westen 1999, Hassin, Uleman & Bargh 2005, Augusto 2010). Therefore, that a part of us has experiences that another part of us has no introspective access to, and therefore cannot report, can be elegantly explained by nested dissociation without any need to postulate anything outside consciousness itself.

Yet, Libet's observations suggest *quick* and *consistent* transitions into awareness of cognition that was initially apparently unconscious. Today we know of a vast variety of analogous cases, such as moving one's foot halfway to the brake pedal before one becomes explicitly aware of danger ahead (Eagleman 2011: 5). The presence of a dissociated 'other' within us all does not seem to explain these cases, for quick and consistent transitions into awareness are not typical of communication between strongly dissociated segments of the psyche. Does this mean that there must exist a neural substrate outside consciousness? No.

Notice that, in order to report an experience—such as making a decision or seeing danger ahead—to another or to oneself, one has to *both* (a) have the experience *and* (b) know *that* one has the experience, which Schooler (2002) called a "re-representation." In other words, one can only report an experience that one is self-reflectively aware of at a metacognitive level. But self-reflection is just a particular, rather complex configuration of consciousness, whereby consciousness turns in upon itself so to experience knowledge *of* its own experiences (Kastrup 2014: 104-110). Nothing precludes the possibility that an experience takes place outside the field of self-reflection without ceasing to be experienced. The latter is still in consciousness, but we cannot report it—not even to ourselves—because we don't know *that* we experience it. Nixon calls it "unconscious experience" (2010: 216), which in my view is an oxymoron, but illustrates the subtlety of the point. And indeed, the existence of experiences that escape metacognitive re-representation is well established in neuroscience today (Tsuchiya et al. 2015, Vandenbroucke et al. 2014).

So the hypothesis here is that *all* mentation is actually conscious, even though we cannot report on much of it—not even to ourselves (Kastrup 2017d[11]). I thus posit that Libet's subjects made their decisions *in consciousness*, but outside the field of self-reflection. Their decisions were 'experiences that weren't aware of themselves' until entering the field of self-reflection after a small delay. Only then could they be reported by the subjects.

In conclusion, the first objection is based on a misunderstanding of terminology. While I use the word 'consciousness' in its broadest sense—that which experiences—the objection assumes it to mean only self-reflective awareness, a particular configuration of consciousness. So it is true that there are phenomena that unfold outside self-reflective awareness, but this does not imply that there is anything outside consciousness itself.

The second objection relates to Explanation 1: I have stated that an organism's inner experiences—including the internally dissociated and non-self-reflective types discussed above—

---

11 This reference can be found in Chapter 9 of the present volume.

cause its body, not the other way around (see also Kastrup 2015: 17-18 & 189-190). However, it is a well-established fact that physical interference with the brain—through psychoactive drugs, trauma to the head, exposure to electromagnetic fields, etc.—affects inner experience. So the arrow of causation must point the other way: from the body to inner experience—or so the objection goes.

Although this may sound persuasive at first, it's based on an unexamined assumption: that the physical is in some sense distinct from, yet causally effective upon, the experiential. This is precisely what I am denying. The proposed ontology asserts that, in essence, there is *only* the experiential, since there is only TWE. The physical is simply the verifiable contents of perception, a particular category of experience. As such, what we call 'physical interference with the brain' is simply the extrinsic appearance of experiential activity external to an alter that, in turn, disrupts the inner experiences of the alter from across its dissociative boundary. The disruption 'pierces through the boundary,' so to speak. The resulting effects are then simultaneously detectable in the extrinsic appearance of the alter—that is, its body—just as the proposed ontology explains. That certain types of experiential activity disrupt other types of experiential activity is not only entailed by the proposed ontology, but is also empirically trivial. After all, our thoughts disrupt our emotions—and vice-versa—every day. For the same reason that thoughts disrupt emotions, 'physical interference with the brain' disrupts an organism's inner experiences.

The third and final objection relates to Explanation 2: if the world we perceive around ourselves is a coded representation of thoughts in universal consciousness, how can the laws of nature be so stable, self-consistent and predictable? The fallacy here, of course, is that of anthropomorphization: to attribute to universal consciousness as a whole cognitive characteristics known only in infinitesimally small dissociated segments of it, such as human beings. Nothing in the proposed ontology precludes the possibility that thoughts in universal consciousness unfold according to very stable, self-consistent and predictable patterns and regularities, whose extrinsic appearance corresponds to the laws of nature. That our human thoughts seem rather reactive and unstable should be considered a product of the evolution of biology, under the pressures of natural selection, within a particular planetary ecosystem. At a universal level, 'that which experiences' has not undergone such evolutionary pressures or processes.

The stability of the laws of nature under the proposed ontology can perhaps be better understood with a simple shift in terminology. Certain schools of psychology speak of "psychological archetypes": innate, built-in templates according to which the dizzying variety of human mentation unfolds (Jung 1991). We can then say that, under the proposed ontology, the laws of nature are akin to the archetypes of universal consciousness. They are built-in templates according to which the dizzying variety of the 'vibrations' of universal consciousness—that is, experiences—develops. As such, the archetypes are analogous to the physical constraints of a vibrating surface, which determine the surface's natural modes of excitation. Although the vibrations themselves can be highly complex and sometimes apparently disorderly, the underlying patterns and regularities remain stable and orderly.

5.8 Comparison to physicalism

How does the proposed idealist ontology measure up to physicalism in terms of parsimony and explanatory power? To begin with, notice that our only access to a world allegedly independent of experience takes the form of perceptions, which are themselves experiences. Therefore, physicalism is an abstract explanatory model produced by thought, not an observable empirical fact. Its motivation is to provide a tentative explanation for Facts 1 to 4.

Physicalism is inflationary: in addition to experience itself—the one undeniable ontological class—it postulates the existence of stuff outside and independent of experience. This step would only be justifiable if we could not make sense of Facts 1 to 4 without it. However, in Section 5.6 we have done just that. Therefore, physicalism can be rejected on grounds of parsimony.

Moreover, physicalism is limited in its explanatory power. *In addition to its own ontological primitives*, it fails to explain experience itself (recall the 'hard problem of consciousness'), which ultimately is all we have. The idealist ontology proposed here, on the other hand, circumvents the 'hard problem' altogether: since consciousness itself is taken to be the sole ontological primitive, it does not need to be explained in terms of anything else. Indeed, from the perspective of the proposed ontology, the 'hard problem' exists only in the logico-conceptual structure of physicalism, with no grounding in empirical reality. By conceptualizing abstractions of consciousness as ontological primitives, physicalists conjure up the impossible challenge of having to reduce consciousness to consciousness's own abstractions.

Since the proposed ontology circumvents the 'hard problem of consciousness' in the process of explaining Facts 1 to 4, physicalism can be further rejected on grounds of explanatory power.

5.9 Comparison to bottom-up panpsychism

By taking complete living beings to be unitary—instead of composite—experiencing subjects (Inference 5), the idealist ontology proposed here avoids the so-called 'subject combination problem' that plagues bottom-up panpsychism (Chalmers 2016). As we've seen, bottom-up panpsychists posit that entities as small as subatomic particles are experiencing subjects in their own merit. They imagine that the unitary subjectivity of more complex experiencing subjects, such as human beings, arises from *bottom-up combination* of countless simpler subjects. The problem is that the bottom-up combination of subjects is an unexplainable process, perhaps incoherent (Coleman 2014). It is just as hard as the 'hard problem of consciousness' (Goff 2009). Inference 5 circumvents this altogether by positing that *top-down dissociation*—instead of bottom-up combination—happens exactly at the level of individual living creatures with unitary subjectivity, such as ourselves. And unlike bottom-up combination, we actually understand and have plenty of empirical evidence for top-down dissociation, as discussed in the context of Inference 4.

The motivation for bottom-up panpsychism is that, undeniably, subatomic particles are the discernible 'pixels' of the world we perceive around ourselves (Explanation 4). But to imagine, for this reason, that the unitary subjectivity of living beings is composed of myriad subatomic-level subjects entails a flawed logical bridge: it attributes to *that which experiences* a structure discernible only in *the experience itself*; that is, in our perceptions of the world. In the framework of the proposed ontology, the error is that of attributing to TWE a structure discernible only in *the excitations of* TWE (Inference 1). This is analogous to saying, for instance, that water is made of ripples simply because one can discern individual ripples in water. Clearly, individual ripples make up the structure of the *movements of* water, not of water itself. Analogously, subatomic particles are the 'pixels' of the observable 'movements' of TWE, not the building blocks of TWE itself. Our unitary subjectivity is not necessarily composed of myriad subatomic-level subjects for the same reason that water is not made of ripples.

Because experiences are excitations of TWE, the latter—by definition—cannot be experienced in its *un*excited state. In the absence of excitations, TWE consists purely in the *potential* for experiences. This is analogous to the 'vacuum state' in Quantum

Field Theory, or to unexcited 'branes' and 'superstrings' in M-Theory and Superstring Theory, respectively. So an argument could be made here that, while water without ripples can be empirically observed to exist, unexcited TWE cannot and, thus, may not exist. Nonetheless, even if TWE exists only in excited states, it remains a conceptual error to conflate patterns of excitation with that which is excited.

5.10 Volition and natural law
The ontology articulated in this paper entails that all reality unfolds in a form of transpersonal consciousness. This tempts some to conclude that, as an implication of the proposed ontology, natural phenomena must be triggered by rationalized volitional choices analogous to our own. One may ask, for instance, *why* universal consciousness has *chosen* to dissociate itself into alters so prone to suffering.

The thinking underlying this question unnecessarily attributes to the whole of universal consciousness particular configurations—such as self-reflection—that it may not have outside its alters. Indeed, I have posited that self-reflection is tied to alter formation (Kastrup 2014: 110-116). As such, the dynamics of universal consciousness outside its alters may unfold, at least partly, along lines that we may describe as instinctual or naturalistic. Moreover, insofar as the question presupposes a form of free will distinct from both randomness and determinism, it may be a red herring (Kastrup 2015: 171-180). The idea that reality unfolds within universal consciousness is not at all incompatible with causes and constraints given by the laws of nature.

5.11 Implications for artificial consciousness
According to the proposed ontology, consciousness is an ontological primitive. As such, it cannot—and does not need to—be created, for it already underlies all nature. Creating

something means inducing a certain pattern of excitation of and in TWE. Universal consciousness is thus that within which all creation happens and out of which all creation is made.

Yet, what is ordinarily meant by 'artificial consciousness' in the field of Strong Artificial Intelligence entails more than just the creation of consciousness proper: it entails the engineering of an entity with *separate, private* conscious inner life, akin to yours and mine. In the context of the proposed ontology, this amounts to *artificially inducing dissociation* in universal consciousness, thereby creating an artificial alter of TWE.

Most attempts to realize 'artificial consciousness' center on mimicking the patterns of information flow discernible in biological nervous systems (e.g. Haikonen 2003, 2007). This, however, only captures the *formal*—not the essential—aspects of alters of TWE. In this specific sense, the attempts are akin to what Feynman called "cargo cult science" (1999: 242-243). Indeed, according to the proposed ontology, a functioning biological nervous system is merely the *extrinsic appearance* of an intrinsic view. It does not logically follow that, by mimicking this appearance, the intrinsic view will also be reproduced. Engineering work in this direction may even succeed in creating philosophical zombies whose behavior is indistinguishable from that of living organisms, but there will be nothing it is like to be these zombies *in and of themselves*, for the same reason that—at least as far as you can tell—there is nothing it is like to be an individual neuron in your brain. There is only something it is like to be your brain *as a whole*—that is, you. Analogously, there is only something it is like to be the non-metabolizing universe *as a whole*, zombies being integral parts of its extrinsic appearance just as the individual neuron is an integral part of your brain.

If biology is the extrinsic appearance of alters of TWE, then the quest for artificial consciousness boils down to abiogenesis: the artificial creation of biology from inanimate matter. If this quest succeeds, the result will again be *biology*, not computer simulations thereof. The differences between flipping microelectronic switches and actual metabolism are hard to overemphasize. Therefore, there is no empirical reason to believe that a collection of flipping switches could ever be what individualized, private conscious inner life looks like from the outside, even if these flipping switches perfectly mimic the patterns of information flow discernible in metabolism.

5.12 Conclusions

I have argued for a coherent idealist ontology that explains reality in a more parsimonious and empirically rigorous manner than mainstream physicalism and bottom-up panpsychism. This idealist ontology also offers more explanatory power than both physicalism and bottom-up panpsychism, in that it does not fall prey to either the 'hard problem of consciousness' or the 'subject combination problem,' respectively. It can be summarized as follows: there is only universal consciousness. We, as well as all other living organisms, are but dissociated alters of universal consciousness, surrounded like islands by the ocean of its thoughts. The inanimate universe we see around us is the extrinsic appearance of these thoughts. The living organisms we share the world with are the extrinsic appearances of other dissociated alters of universal consciousness. As such, the quest for artificial consciousness boils down to the quest for abiogenesis. The currently prevailing concept of a physical world independent of consciousness is an unnecessary and problematic intellectual abstraction.

Chapter 6

Making sense of the mental universe

This article first appeared in *Philosophy and Cosmology*, ISSN: 2518-1866 (online) and 2307-3705 (print), Vol. 19, pp. 33-49, on 26 September 2017. *Philosophy and Cosmology*—published since 2004 by the International Society of Philosophy and Cosmology, Kiev, Ukraine—is one of the very few open-access academic journals in the world to specialize on the interface between ontology and the physical sciences, which is the subject matter of this article. A summary of the present article has also appeared in *Scientific American* on 16 February 2018.[1]

6.1 Abstract

In 2005, an essay was published in *Nature* asserting that the universe is mental and that we must abandon our tendency to conceptualize observations as things. Since then, experiments have confirmed that—as predicted by quantum mechanics—reality is contextual,[2] which contradicts at least intuitive formulations of realism[3] and corroborates the hypothesis

---

1 At the time of this writing, the *Scientific American* essay was freely available online at: https://blogs.scientificamerican.com/observations/thinking-outside-the-quantum-box/.

2 That is, the values of physical properties fundamentally depend on how the properties are observed. This runs counter to physicalist intuition, according to which the mass, charge, spin, position and momentum of a physical entity should be whatever they are regardless of whether and how they are observed.

3 That is, the notion that physical properties exist in some sense independently of mind.

of a mental universe. Yet, to give this hypothesis a coherent rendering, one must explain how a mental universe can—at least in principle—accommodate (a) our experience of ourselves as distinct individual minds sharing a world beyond the control of our volition; and (b) the empirical fact that this world is contextual despite being seemingly shared.[4] By combining a modern formulation of the ontology of idealism with the relational interpretation of quantum mechanics, the present paper attempts to provide a viable explanatory framework for both points. In the process of doing so, the paper also addresses key philosophical qualms of the relational interpretation.

6.2 Introduction

The recent loophole-free verification of Bell's inequalities (Hensen et al. 2015) has shown that no theory based on the joint assumptions of realism and locality is tenable. This already restricts the viability of realism—the view that there is an *objective physical world*; that is, a world (a) ontologically distinct from mentation that (b) exists independently of being observed— to nonlocal hidden-variables theories. More specifically, other recent experiments have shown that the physical world is *contextual*: its measurable physical properties do not exist before being observed (Gröblacher et al. 2007, Lapkiewicz et al. 2011, Manning et al. 2015). Contextuality is a formidable challenge to the viability of realism.[5]

These developments seem to corroborate Richard Conn Henry's assertion in his 2005 *Nature* essay that "The Universe is entirely mental" (Henry 2005: 29). After all, in a mental universe (a) observation necessarily boils down to perceptual experience— what else?—and (b) the physical properties of the world exist only insofar as they are perceptually experienced. There is no ontological ground outside mind where these properties could otherwise reside before being represented in mind. Indeed, in a mental universe observation *is* the physical world—not merely a representation *of* the world—which not only echoes but makes sense of contextuality.

Realism, on the other hand—at least in its intuitive formulations—entails that the world should have *objective physical properties*; that is, properties ontologically distinct from mentation, which exist even without being observed. Accurate observation should simply *reveal* what the objective physical properties of the world already were immediately prior to being observed, which is contradicted by contextuality.

There have been attempts to preserve some form of realism by finding a subset of physical properties whose values can be determined in a non-contextual manner under certain circumstances. The idea is then to claim that this subset *is* the objective physical world. For instance, Philippe Grangier (2001), inspired by Einstein-Podolsky-Rosen's view (1935) of what constitutes physical objectivity, contends that the quantum state of a system, defined "by the values of a set of physical quantities, which can be predicted with certainty and measured repeatedly without perturbing in any way the system" (Grangier 2001: 1), is an objective physical entity.

The problem with this approach is highlighted by Grangier himself: the "definition [of the quantum state] is inferred from

---

4 After all, if the world isn't standalone but, instead, fundamentally depends on observation, the obvious question is why we all seem to observe the same world.

5 An extensive and more explicit discussion of what contextuality is, why it contradicts realism and how it has been experimentally verified over the past couple of decades, is included in Chapter 15, Section 15.3 of this volume. Unlike the brief summary here, the discussion in Chapter 15 is meant to be accessible to non-physicists and uses layman's terms. It may be useful for interested readers without a physics background to read Section 15.3—which is a self-contained section—at this point, before continuing on with the remainder of this chapter.

observations which are made at the macroscopic level" (Ibid.: 2). In other words, the supposedly physically objective quantum state of a system depends on the *a priori* existence of a physically objective classical world surrounding the system. This begs the question of physical objectivity instead of rendering it viable under contextuality. Because "a quantum state 'involving the environment' cannot be consistently defined" (Ibid.: 4), Grangier's approach fails to reconcile contextuality with a supposedly physically objective world.

Some nonlocal hidden variables theories that preserve non-intuitive forms of realism—such as perhaps Bohm's (1952a, 1952b)—may still be reconcilable with contextuality. However, these theories postulate—often at the cost of mathematical acrobatics—extra theoretical entities that are both empirically ungrounded and unnecessary for predictive purposes.

Carlo Rovelli's *relational interpretation* (2008), on the other hand, sticks to plain quantum theory and embraces contextuality. Instead of loading it with unnecessary baggage, it simply interprets what quantum theory tells us about the world and bites the bullet of its implications. Rovelli's goal "is not to modify quantum mechanics to make it consistent with [his] view of the world, but to modify [his] view of the world to make it consistent with quantum mechanics" (Ibid.: 16).

In the remainder of this paper, I shall take the relational interpretation as my working hypothesis. My motivation for doing so is three-fold: (a) the interpretation is consistent with experimentally-verified contextuality; (b) it is parsimonious in that it does not postulate predictively-redundant hidden variables; and (c) Rovelli's case for why other approaches are inferior to the relational interpretation is compelling (Ibid.: 16-19).

By embracing contextuality, the relational interpretation regards every property of the physical world as relative to the observer. This is analogous to how the speed of a particle with mass is always relative to its observer. *There are no absolute physical quantities*, but simply a set of relational properties that comes into existence depending on the context of observation. Rovelli summarizes it thus:

> If different observers give different accounts of the same sequence of events, then each quantum mechanical description has to be understood as relative to a particular observer. Thus, a quantum mechanical description of a certain system (state and/or values of physical quantities) cannot be taken as an "absolute" (observer independent) description of reality, but rather as a formalization, or codification, of properties of a system relative to a given observer. Quantum mechanics can therefore be viewed as a theory about the states of systems and values of physical quantities relative to other systems. (Ibid.: 6)

Like the Copenhagen interpretation, the relational interpretation entails that (a) physical quantities are products of observation. But most significantly, it goes further than Copenhagen by asserting that (b) the world is *relational*: an observation does not create a world shared by everyone, but just the world of that particular observer.

This difference with respect to the Copenhagen interpretation is not trivial. After all, it is implausible but conceivable that observation could create an objective physical world shared by all observers. For instance, if never observed, the spin of an electron may lack physical objectivity. But its *first* observation would then, *ex hypothesi*, determine its physical value for all subsequent observers. The physical objectivity of this value — and thus of the world — could be inferred from *consensus* among these observers. Such a hypothesis is consistent with assertion (a) above but not (b).

It is also conceivable that each of us could be living alone in an objective physical world—that is, a physical world ontologically distinct and independent from our mentation—peculiar to ourselves. The physical objectivity of such a world could be inferred from *non-contextuality* verified by experiment. Such a hypothesis is consistent with assertion (b) above but not (a).

By combining assertions (a) and (b), the relational interpretation renders realism—the notion that there is an objective physical world—meaningless. After all, in the absence of consensus and non-contextuality, on the basis of what could we speak of physical objectivity? What meaning would the latter have? According to the relational interpretation, the world exists only insofar as the information associated with an observer is concerned.

Rovelli seeks to avoid ontological conjectures. Yet, the denial of realism seems to be a direct implication of the relational interpretation. In fact, it is only one among a handful of philosophical issues Rovelli admittedly leaves unaddressed: "I am aware of the 'philosophical qualm' that the ideas presented here may ... generate," he writes. "I certainly do not want to venture into philosophical terrains, and I leave this aspect of the discussion to competent thinkers" (Ibid.: 19).

It is these philosophical qualms that the present paper attempts to tackle, without contradicting the relational interpretation. In doing so, it articulates a mind-only idealist framework consistent with contextuality and—contradictory as this may at first sound—our intuition that we are individual beings sharing experiences with each other.

6.3 First qualm: The intuition of a shared world
The relational interpretation denies that we can all inhabit the same objective physical world. It implies instead that each of us—as different observers—lives alone in our own private physical world, created according to the context of our own private observations. Insofar as this resembles metaphysical solipsism,

it may be philosophically problematic. However, there still is a way to uphold our intuition that there is a consensus reality we share with other people.

It is true that, according to the relational interpretation, observation is not a measurement *of* or *in* a shared physical world, but the process that brings a unique physical world into existence in relation to each particular observer. This way, there are as many physical worlds as there are observers. A way to visualize this is to imagine that each person sits alone in a car corresponding to his or her own physical world. No two people can ever sit in the same car. Any ontology that contradicts this is inconsistent with the relational interpretation.

However, we can still ask another question: *Can the physical worlds of different observers be consistent with, and similar to, each other?* Notice that this does not deny that different observers have their own physical worlds; it simply asks whether these distinct worlds can be *similar* or *mutually consistent*. In other words, the question is whether we could all be sitting in cars of the same make, model and year; cars that, although *distinct*, are nearly indistinguishable from each other. If so, we would each describe our own cars in a way consistent with all other descriptions.

Here is what the relational interpretation has to say about this: nothing precludes the *possibility* of the physical worlds of different observers being similar or mutually consistent. However, it is fundamentally impossible to assert that they *are* so, for "the information possessed by distinct observers cannot be compared directly" (Ibid.: 14). The rationale here is as follows: the notion of a consensus physical reality emerges from inter-personal communication. If I stand on a beach watching the waves and the person next to me also reports seeing waves, it is this inter-personal communication that leads me to believe that I and the other person experience the same beach. However, what I hear the other person say is itself the result of *my* observation, which brings the other person's report into existence *in relation to me.* As such, the other person's report is itself as aspect of *my* physical world as a particular observer; it has no absolute existence. For all I know, the physical worlds experienced by other people—as *distinct* observers who bring *distinct* physical worlds into existence—may be entirely different from mine. The consensus I believe to exist about external reality may itself be an element peculiar to *my* physical world, my car. Everybody is an observer locked in his or her own car. There is no privileged referee who could walk from car to car, collect and compare the descriptions of each car, and then verify whether there actually is a consensus.

All this said, the intuition of a consensus external reality is so strong that we must ask: Can there be an ontological underpinning for the relational interpretation whereby the respective physical worlds of different observers are at least *expected* to be similar or mutually consistent? In other words, can an ontology provide us *good reasons to believe*—even though we fundamentally could never verify it—that the physical worlds of different observers *should* look alike? The motivation for this question is admittedly subjective, but the exact same subjective motivation has been enough to marginalize metaphysical solipsism throughout the history of philosophy. Indeed, Bertrand Russell's argument against solipsism seems to be applicable here: the idea that we might each be alone in an idiosyncratic world of our own "is psychologically impossible to believe, and is rejected in fact even by those who mean to accept it" (Russell 2009: 161).

I shall shortly attempt to articulate an idealist framework for the relational interpretation according to which similarity or consistency across physical worlds is the natural and expected case, even though it cannot be verified. This framework is meant to acknowledge and assuage the intuition that we share the experiences of life with other people, whilst upholding contextuality.

6.4 Second qualm: The ontological ground of information

The relational interpretation relies on Shannon's concept of information: "A *complete* description of the world is exhausted by the relevant *information* that systems have about each other" (Rovelli 2008: 7, emphasis added). Although Rovelli avoids explicit ontological commitments, his appeal to information according to Shannon's definition (1948) implies one such commitment. After all, Shannon defines information as a measure of the number of states discernible in a system. As such, information is an abstraction associated with the possible *configurations of* a system, not a thing unto itself (unless, of course, one is prepared to venture into the abstraction wilderness of ontic pancomputationalism (Fredkin 2003, Tegmark 2014)). Hence, insofar as it relies on (Shannon) information, the relational interpretation requires either a realist world—wherein information is grounded in the discernible states of objective physical arrangements—or an idealist world—wherein information is associated with the discernible qualities of experience. And since realism is meaningless under the relational interpretation, idealism seems to be implied by it.

However, idealism faces some challenges. In another work (Kastrup 2017c[6]), I have addressed and hopefully refuted common objections to it. In this paper, two challenges will be more thoroughly looked at: If mind extends into the world itself, grounding it ontologically, why can we not mentally control or at least influence the laws of physics? Moreover, if all reality is mental, then there is no non-mental stuff to insulate different individual minds from one another. Why, then, can we not directly access each other's thoughts? Satisfactorily answering these challenges is another key objective of the idealist framework I shall attempt to articulate shortly. If successful, the articulation

---

6 This reference can be found in Chapter 8 of the present volume.

will render idealism a viable ontological underpinning for the notion of (Shannon) information intrinsic to the relational interpretation.

6.5 Third qualm: Relationships without absolutes

The central idea of the relational interpretation is the notion that "physics is fully relational, not just as far as the notions of rest and motion are considered, but with respect to *all* physical quantities" (Rovelli 2008: 7, emphasis added). The problem here is that the analogy with rest and motion, albeit intuitively appealing, breaks when applied to "all physical quantities."

Indeed, the relational nature of rest and motion depends on certain posited absolutes, such as defined particles. To say, for instance, that the speed of particle A is one with respect to particle B and another with respect to particle C is conditioned upon the existence of particles A, B and C as non-relational entities. Rest and motion have meaning only insofar as they are relationships between absolutes. But if *all* physical quantities are to be regarded as relational, what absolutes give these relationships meaning? To speak of relationships between relationships immediately implies infinite regress.

Let us take a step back. What the relational interpretation actually requires is that all *physical* quantities be relational. As such, it would only imply infinite regress if physical quantities were all there is. On the other hand, if an ontological underpinning for the relational interpretation could accommodate absolutes that are *not* physical quantities, infinite regress could be avoided. This is what the idealist framework ahead also does, as I shall soon elaborate upon.

Notice that, although positing absolutes that are not physical quantities is necessarily a *meta*physical step, it is not empirically ungrounded. There is an empirically-accessible ontological ground where absolutes can be found that are not—unless *assumed* or, at best, *inferred* to be so on philosophical grounds—

physical quantities: mind and its thoughts. I shall elaborate further on this claim shortly.

6.6 Fourth qualm: The meaning of 'physical world'

When we speak of a 'physical world' we often make implicit ontological assumptions about it, such as non-contextuality and realism. However, as we have seen, these assumptions are stripped off by the relational interpretation. So what does 'physical world' mean under it?

The clarity of thought of Andrei Linde comes to our aid at this point:

> Let us remember that our knowledge of the world begins not with matter but with perceptions. ... Later we find out that our perceptions obey some laws, which can be most conveniently formulated if we assume that there is some underlying reality beyond our perceptions. This model of material world obeying laws of physics is so successful that soon we forget about our starting point and say that matter is the only reality, and perceptions are only helpful for its description. This assumption is almost as natural (and maybe as false) as our previous assumption that space is only a mathematical tool for the description of matter. (1998: 12)

So, in the absence of non-contextuality and realism, the 'physical world' of the relational interpretation can only be the *contents of perception*. There is nothing else the physical world could be. Indeed, as Linde pointed out, physics ultimately pertains to the study of the patterns and regularities of perception. As such, the "physical quantities" referred to by Rovelli are part of the contents of perception.

It could be argued at this point that quantum phenomena occur at a microscopic scale that cannot be perceived *directly*, but only through instrumentation. Yet, even in this case, whatever we know about these microscopic quantum phenomena is still a part of the contents of perception: physicists *perceive* the output of instrumentation. When predicting microscopic quantum behavior, physicists are in fact predicting the perceivable output of instrumentation. *Physics is entirely about what is perceived,* directly or indirectly.

We know that next to the physical world—that is, next to the contents of perception—there are also non-perceptual mental categories such as thoughts (for simplicity, I shall henceforth refer to all non-perceptual mental categories simply as 'thoughts'). Many physicists posit that thoughts should be explainable in terms of physical quantities and, as such, become part of the physical world *by reduction*. However, this is a philosophical *assumption* that does not change the scientific fact that quantum mechanics does *not* predict thoughts; it only predicts the unfolding of perception. In the absence of non-contextuality and realism, the physical world of quantum mechanics *is* perception.

Attentive readers will have noticed that I have just opened a door for tackling the third qualm of the relational interpretation, as discussed in the previous section. More on this shortly.

6.7 Fifth qualm: The meaning of 'physical system'
Under the relational interpretation, all "physical systems" are valid observers and can, in turn, also be observed (Rovelli 2008: 4). This neutrality is a strength, for it circumvents a host of issues regarding what constitutes an observer. Yet, the same neutrality disguises the fact that a deeper question is left unanswered: *What constitutes a physical system to begin with?* From a philosophical perspective, the answer is not self-evident.

The intuition behind what we ordinarily regard as discrete physical systems—such as tabletop measurement apparatuses—entails (a) delineating a subset of the physical world on structural or functional grounds and (b) treating this subset as an entity in some sense separate from the rest of the physical world. The question is whether such delineation is ontic or epistemic.

If the delineation merely helps us structure our *conceptual knowledge of* the physical world, it is epistemic and—despite being convenient—arbitrary on an ontic level. For instance, although the handle of a mug is cognitively salient and can be conveniently treated as a separate entity, distinguishing it from the mug is arbitrary. Another example: the delineation of what we call a car, though based on structural and functional reasoning, is arbitrary. If I argue that, say, the spark plugs are integral to the car because without them the car cannot function, by the same token I would also have to include the fuel that makes its engine run, the environment air that allows combustion and cools the engine, the road gripped by the tires, the ground that sustains the road, the gravity that enables grip, and so on. The decision of where to stop is merely epistemic, motivated by convenience. As such, epistemic delineation is somewhat akin to finding faces in clouds or tracing figures on tree bark. Subsets of the physical world traced in this manner exist only conceptually and, therefore, cannot be considered proper physical systems in their own merit. There is no physical reason to carve them out of the context within which they were traced.

A proper physical system must be an internally integrated whole separate, in some *ontic* sense, from the rest of the physical world. The problem is that there are strong reasons—largely based on quantum mechanics itself—to think that the *entire universe* is one integrated whole without ultimate parts. Jonathan Schaffer, for instance, points out that

> physically, there is good evidence that the cosmos forms an entangled system and good reason to treat entangled systems as irreducible wholes. Modally, mereology allows for the possibility of *atomless gunk*, with no ultimate parts for the pluralist to invoke as the ground of being. (2010: 32, original emphasis.)

Horgan and Potrč (2000) have also argued that only the universe as a whole can be considered a concrete entity in its own merit, which they called the "blobject." Thus, only the "blobject" can be a proper physical system.

What this line of thought suggests is that no subset of the physical world can be considered a proper physical system; only the physical world *as a whole* can. Everything we regard as subsystems of the physical world—such as tabletop measurement apparatuses—arises from epistemic delineation and is, in a sense, akin to figures traced on tree bark. The physical substantiation of this conclusion is von Neumann's chains (von Neumann 1996): 'subsystems' of the inanimate world never perform measurements, but simply become entangled with each other upon interacting.

If only the universe as a whole were a proper physical system, observ*er* and observ*ed* would be the same system, leading to untreatable self-reference. But I contend that there is precisely *one* criterion of delineation that is *not* arbitrary, and by virtue of which we can ontologically ground observ*er* and observ*ed* without self-reference. I shall discuss this in the next section, wherein I begin to elaborate on my proposed idealist framework.

6.8 Mind and alters

Here is a brief recapitulation of the preceding sections: (a) the relational interpretation renders the notion of realism meaningless. Therefore, (b) it implies idealism insofar as it relies on (Shannon) information, and (c) it implicitly defines the physical world as the contents of perception. However, it then raises problems such as (d) why we cannot mentally influence the laws of nature if mind extends into the physical world, (e) how distinct individual minds can exist in a fully mental universe, (f) what constitutes a proper physical system, and (g) how relationships can exist without absolutes to give them meaning and avoid infinite regress. Despite its relational character, the interpretation still (h) leaves a door open for the intuitively appealing possibility of similar or mutually consistent physical worlds across observers.

From (b), the most direct and parsimonious ontological underpinning for the relational interpretation is that *reality consists of patterns of excitation of a universal mind*. This is analogous to how quantum field theory posits that reality consists of patterns of excitation of a quantum field and M-theory those of a hyper-dimensional 'brane.' The universal mind could, in principle, even accommodate the same mathematical formalisms of these theories. The main difference is that, unlike the quantum field and the hyper-dimensional 'brane,' the universal mind is not an *object* but the *subject*. Its excitations are thus *experiences* — not objective physical values — though they can still be *modeled* or *described* mathematically by values.

Let me dwell a little longer on this move, for it is profoundly counterintuitive to most physicists and even philosophers. Every theory of nature must rely on at least one ontological primitive, since we cannot keep on explaining one thing in terms of another forever. According to quantum field theory, the quantum field is the ontological primitive, so everything about the physical world should be reducible to excitations of the quantum field. What I am proposing here is that *universal mind* is the ontological primitive. An alternative but equivalent way of saying the same thing is to say that the quantum field *is mind*; that is, the subject, not an abstract conceptual object. Inferring universal mind — not your or my *personal psyche*, mind you, but mind as a generic ontological class — to be nature's sole ontological primitive is thus a valid conjecture, at least in principle. It is equally valid to think of the dynamics of nature as being constituted by the *excitations of* universal mind — that is, experiences themselves, with all the entailed *qualities* — much as quantum field theory thinks of the dynamics of nature as excitations of an abstract quantum field and M-theory those of an abstract brane. The challenge, of course, is to explain the patterns and regularities of nature in terms of subjectivity alone.

Indeed, the idea that a universal mind is nature's single ontological primitive immediately brings problem (e) to the forefront. As it turns out, the solution of this problem is the same as that of problem (f), so I will start my argument by addressing the latter.

Rovelli "assume[s] that the world can be decomposed (possibly in a variety of ways) in a collection of systems, each of which can be equivalently considered as an observing system or as an observed system" (2008: 10). The problem, as we have seen in the previous section, is that the criteria for this decomposition seem arbitrary both from ontic and physical perspectives.

There is, however, one very natural ontic decomposition. To see it, notice that the boundaries of our own body are not arbitrary. Our ability to *perceive* ends at the surface of the body: our skin, retinas, eardrums, tongue and the mucous lining of our nose. We cannot perceive photons hitting a wall or air pressure oscillations bouncing off a window, but we *can* perceive those impinging on our retinas and eardrums, respectively. Moreover, our ability to act through direct intention also ends at the surface of the body: we can move our arms and legs simply by *intending* to move them. However, we cannot do the same with tables and chairs. Clearly, thus, the delineation of our body is not a question of epistemic convenience: it is an *empirical fact*. I cannot just decide that the chair I am sitting on is integral to my body, in the way that I *can* decide that the handle is integral to the mug. Neither can I decide that a patch of my skin is not integral to my body, in the way that I *can* decide that the hood is not integral to the jacket. The criterion here is not merely a functional or structural one, but *the range of mentation — sensory perception, intention — intrinsically associated with our body*. Based on this ontic criterion, there is no epistemic freedom to move boundaries at will.

Insofar as we can assume that all living creatures have mental life and inanimate objects do *not*, this conclusion can be generalized as follows: *living bodies are proper physical systems*; they *can* be carved out of their context. Therefore, only the *inanimate* universe *as a whole*—that is, one universal von Neumann chain—and individual living bodies are proper physical systems; only the inanimate universe and living bodies are observers. Everything else is akin to figures traced on tree bark.

Now, since the ontic criterion for delineating bodies is the range of mentation associated with them, *each proper physical system is associated with its own bounded mentation*. Yet, how can bounded mentation exist within one universal mind?

The answer is empirically motivated: it is now well-established in psychiatry that mental contents can become *dissociated* (Kelly et al. 2009: 167-174, Schlumpf et al. 2014, Strasburger & Waldvogel 2015); that is, undergo "a disruption of and/or discontinuity in [their] normal integration" (Black & Grant 2014: 191). Indeed, the normal integration of mental contents takes place through chains of *cognitive associations*: a perception may evoke an abstract idea, which may trigger a memory, which may inspire a thought, etc. These associations are *logical*, in the sense that e.g. the memory inspires the thought because of a certain *implicit logic* linking the two. Integrated mentation can then be modeled by a connected directed graph, as shown in Figure 6.1a. Each vertex represents a particular mental content and each edge a cognitive association logically linking mental contents together. Every mental content in the graph of Figure 6.1a can be reached from any other mental content through a chain of cognitive associations. Dissociation, in turn, occurs when the graph becomes disconnected, as shown in Figure 6.1b. Some mental contents can then no longer be reached from others. Following the psychiatric convention, I shall refer to the subgraph with grey vertices as a (dissociated) *alter*.

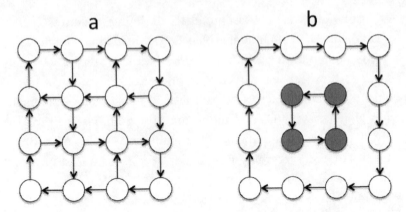

Figure 6.1: A connected graph (a) illustrating integration of mental contents, and a disconnected graph (b) illustrating dissociation and the corresponding formation of an alter (subgraph in grey).

Because cognitive associations are essentially logical, as opposed to spatio-temporal, the scheme of representation in Figure 6.1 allows for the *simultaneous* experience of multiple mental contents linked together in a connected subgraph. This is empirically justifiable: a perception, for instance, can be experienced at the same time as the thoughts it evokes and the emotions evoked by these thoughts. Moreover—and by the same token—the two disconnected subgraphs in Figure 6.1b can also represent two *concurrently conscious* dissociated subjects of experience. The motivation for this is again empirical: there is compelling evidence that different alters of the same psyche can be co-conscious (Kelly et al. 2009: 317-322, Braude 1995: 67-68).

An alter loses direct access to mental contents surrounding it, *but remains integral to the mind that hosts it.* The disconnection between an alter and surrounding mental contents is merely logical. As an analogy, a database may contain entries that are not indexed and, therefore, cannot be reached, but this does not physically separate those entries from the rest of the database.

Dissociation elegantly explains how mental processes can become bounded and disconnected from each other without the need to invoke anything ontologically distinct from mind. As such, dissociation is at least a useful analogy to explain how distinct individual psyches can form under idealism: *each is an alter of universal mind.* A living body is simply the extrinsic appearance of an alter (more on this shortly). Dissociation thus solves not only problem (e), but also (d): we, as alters of universal mind, cannot mentally influence the laws of nature because we are dissociated—logically disconnected—from the corresponding mental contents.

At this point, the reader may feel tempted to explain away dissociation in terms of information flows in physically objective brain tissue. Notice, however, that this would assume realism and beg the very question of ontology being addressed here. The hypothesis I am just beginning to elaborate on is precisely that dissociation—as a phenomenon that *precedes* the physical world ontologically—can *explain* the physical world, as opposed to the other way around. This is somewhat analogous to saying, under quantum field theory, for instance, that certain fundamental excitatory phenomena of the quantum field give rise to the physical world we can measure. So please bear with me.

6.9 Mental impingement across a dissociative boundary

By definition, mental contents inside an alter cannot directly evoke mental contents outside the alter, and vice-versa. But they can still *influence* each other. Indeed, mental impingement across a dissociative boundary is empirically known. Lynch and Kilmartin, for instance, report that dissociated feelings can dramatically affect our thoughts (2013: 100), while Eagleman shows that dissociated expectations routinely mold our perceptions (2011: 20-54). We can visualize this as in Figure 6.2a, wherein the partial overlap of adjacent vertices internal and external to the alter represents impingement across the dissociative boundary.

Figure 6.2b illustrates the same thing according to a simplified representation unrelated to graph theory: the broader mind is represented as a white circle with an alter represented as a grey circle within it. These circles are no longer graph vertices but represent sets of mental contents. The dashed arrows represent the impingement of external and internal mental contents—not explicitly shown—on each other, across the alter's boundary. For clarity, notice that these dashed arrows do *not* represent cognitive associations. I shall use this simplified representation henceforth.

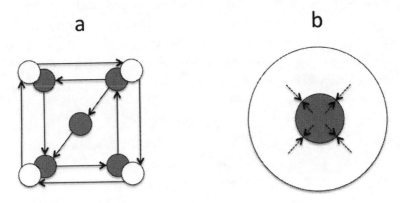

a   **b**

Figure 6.2: Mental contents impinging on the dissociative boundary of an alter, illustrated in two different but equivalent ways, (a) and (b).

6.10 A physical world as Markov Blanket

If—as idealism posits—a universal mind is the single ontological primitive underlying all nature, then the formation of an alter defines a boundary within this mind that separates mental contents enclosed by the boundary from mental contents outside the boundary. Now, as we have seen in the previous section, certain mental contents within and outside the boundary can also impinge on each other (Figure 6.2). Three different types of mental state can then be defined with respect to an alter: internal, external and interactive state, the latter resulting from impingement. The boundary of an alter is thus akin to a Markov Blanket (Pearl 1988). For this reason, and inspired by Friston's model (2013), I shall represent the interaction of an alter with its surrounding mental environment as illustrated in Figure 6.3.

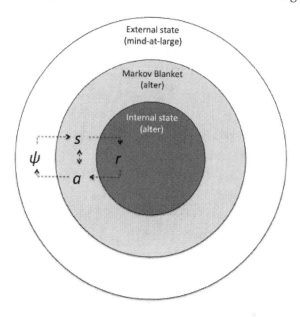

Figure 6.3: An alter—with internal mental state r—interacting with the surrounding environment—with external mental state ψ—through the sensory (s) and active (a) states of its Markov Blanket.

I shall refer to the segment of universal mind that is not comprised in any alter as 'mind-at-large.' Therefore, the state ψ of mind-at-large is external to all alters. An alter, in turn, has internal state r and interacts with mind-at-large through sensory state s and active state a, both comprised in its Markov Blanket. Sensory state s represents the (Shannon) information the alter has about its surrounding mental environment. Active state a represents the alter's manifest intent: mental action that perturbs the environment. Notice that 'environment' here does *not* refer to the physical world, but to *non-physical* thoughts surrounding the alter. I shall further clarify this shortly.

Sensory state s depends on external state ψ and action state a. The dependency of s on a, however, is both indirect—operating through the influence of a on ψ—and direct, as shown in Figure 6.3. The direct dependency represents the quantum mechanical fact that the information an alter—as observer—has about its surrounding environment depends both on the environment itself (ψ) and on how the alter's manifest intent (a) causes it to observe the environment. The classical example is the experimentally confirmed fact that even whole atoms behave either as waves or as particles depending on how the experimenter chooses to observe them (Manning et al. 2015).

The double dependency of sensory state s on active state a can be justified as follows: on the one hand, the manifest intent a of an alter perturbs—through mental impingement across its dissociative boundary, as illustrated in Figure 6.2b—the state ψ of mind-at-large itself. On the other hand, the manifest intent a also determines the specific 'vantage point' the alter has on mind-at-large and, therefore, what information the alter gathers about it. As an analogy, when one holds up a snow globe, this intentional action not only perturbs the state of the snow globe itself, but also determines the vantage point from which one looks at the snow globe.

The active state a depends on internal state r and sensory state s. The dependency of a on s is again both indirect—operating through the influence of s on r—and direct, as shown in Figure 6.3. A simple analogy justifies this double dependency: an alter's manifest intent depends both on the information the alter has about the environment (s) *and* on what the alter *thinks about* this information (r).

External state ψ and internal state r are *thoughts* of mind-at-large and an alter, respectively. I submit that quantum superposition states are *models* of these thoughts, the evolution of the latter being governed by Schrödinger's equation. To gain intuition about this, imagine the following: you have received a job offer but remain undecided about whether to accept it or not. Your thoughts then remain in a form of superposition, encompassing two binary alternatives simultaneously: accepting and refusing the offer. Each alternative is associated with the *degree of affinity* you have with it—which translates into your tendency to choose it—at that particular moment in time. I posit that a quantum superposition is simply a *second-person model* of this type of ambivalent mental state that we all experience from a first-person perspective. External state ψ is a model of what it is like to *be* mind-at-large in the process of entertaining conflicting alternatives concurrently in its imagination. As such, the wave function of ψ does represent epistemic uncertainty; *but—crucially—the epistemic uncertainty of mind-at-large itself, not of the alter observing it.*

I further posit that the process of observation consists in the interaction between external state ψ and internal state r through the Markov Blanket. In this context, it is tempting to simply say that active state a represents the intentional act of observation and sensory state s the outcome of this act. However, as discussed above and illustrated in Figure 6.3, a and s are co-dependent and cannot be teased apart. So a better way to think about the process of observation may be suggested by the following analogy: insofar as ψ and r can both be modeled by a wave function, they can be regarded as *thought waves* encompassing a set of binary alternatives with associated degrees of affinity, just as discussed in the example of a job offer above. Observation can then be modeled as the *interference pattern*—whose compound state is represented in Figure 6.3 by s and a—produced when these thought waves interact with each other within the Markov Blanket. Interference favors one of every pair of superposed alternatives in ψ by amplifying its experience while dampening the other. The result is our perception of a definite, classical world. The alternative favored can be regarded as the common denominator of the affinities embedded in ψ and r.

The interpretation suggested above shall remain a matter of philosophical speculation until somebody writes down the wave function for the thoughts of a conscious human being (r) and formalizes the interaction dynamics between it and ψ. This echoes Zurek's view that

> an exhaustive answer to [the question of why we perceive a definite world] would undoubtedly have to involve *a model of "consciousness,"* since what we are really asking concerns our (observers) impression that "we are conscious" of just one of the alternatives. Such model of consciousness is presently not available. (1994: 29, emphasis added.)

Whatever the case, under the relational interpretation—as discussed earlier—the physical world *is* perception. Therefore, it is determined by sensory state s, which in turn is co-dependent on active state a. The next step in this line of reasoning is inevitable: *the physical world is the Markov Blanket.* Everything else—that is, ψ and r—is non-physical *thought*. It is the interaction between ψ and r that produces perception and, thus, the physical world. Only states s and a of the Markov Blanket are *physical* states, for only they are comprised in the physical world.

Since *all* states in Figure 6.3 are ultimately patterns of excitation of universal mind, physical states represent but a particular class of mental states: namely, perceptual states. Another class is exemplified by states ψ and r, which consist of pure thought. So, while discernible from each other qualitatively, *both physical and non-physical states are ultimately mental.*

6.11 Extrinsic appearances

Figure 6.3 can be extended to multiple alters, as illustrated in Figure 6.4. The interaction between ψ and the internal state r of each alter creates the physical world *of this alter* in the form of its respective Markov Blanket encompassing sensory state s and active state a. Therefore, each alter has its own physical world. The wave function of ψ also becomes quantum mechanically correlated, upon interaction, with the active state a of each alter. This way, ψ accrues (Shannon) information about the presence and behavior of all alters interacting with mind-at-large.

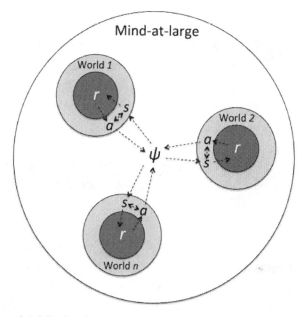

Figure 6.4: Mind-at-large and alters. When an alter interacts with mind-at-large, ψ becomes quantum mechanically correlated with the alter's active state a, so that each alter can indirectly obtain information about all other alters through its sensory state s.

This has a significant implication. When an alter A_1 interacts — through mental impingement across its dissociative boundary — with mind-at-large, ψ becomes quantum mechanically correlated with active state a_1 of A_1. Therefore, when an alter A_2 subsequently interacts with mind-at-large, the information its sensory state s_2 acquires can include information about A_1. The physical world of A_2 can thus reflect the presence and actions of

The Idea of the World

A_1. I shall refer to the information about A_1 in s_2 as A_1's *extrinsic appearance in relation to A_2*.

For reasons discussed in Section 6.8, I posit that the extrinsic appearances of other alters in relation to us are the living bodies we *perceive* around ourselves: other people, animals, possibly plants. As such, biology is what betrays elements of our world as extrinsic appearances of other alters.

This notion of extrinsic appearances can be extended to mind-at-large itself: the *inanimate* universe is the extrinsic appearance of mind-at-large in relation to us. That there is overwhelming evidence for the existence of the universe before the rise of life means solely that the universal mind existed before its first alter ever formed; that is, before abiogenesis.

The extrinsic appearances of other alters in relation to us are aspects of our respective physical world. They constitute proper physical systems within our Markov Blanket. Therefore, *only living beings and the inanimate universe as a whole constitute observers*. All other subsystems of the inanimate universe—such as tabletop measurement apparatuses—are only subsets of 'pixels' integral to mind-at-large's extrinsic appearance. There is no more reason to carve them out as separate subsystems in their own merit than there is reason to carve out the subset of reddish pixels of a photograph and treat it as a thing in its own merit.

At this point, it is important to notice that the external state ψ of mind-at-large and the internal state r of each alter do *not* have spatio-temporal extension, for they are *not* physical. Their seemingly spatial representation in Figure 6.4 is simply an artifact of depiction. Space is confined to Markov Blankets. Everything else is pure thought.

6.12 Consistency with the relational interpretation

Let us now verify, point by point, that the ontological framework discussed in Sections 6.8 to 6.11 is consistent with the relational interpretation:

118

- *The relational interpretation entails that all physical quantities are created by observation.* This is reflected in Figure 6.3, wherein physical quantities are represented by the sensory state s of the alter, which arises only from the interaction—that is, observation—between ψ and r.

- *The relational interpretation asserts that there is no absolute physical world, all physical quantities being relational.* This is reflected in Figure 6.4, wherein each alter—as observer—has its own Markov Blanket, which arises from the alter's own interaction with mind-at-large.

- *The relational interpretation asserts that no physical system is privileged: all physical systems can observe and be observed.* Indeed, mind-at-large and all alters—that is, all physical systems—can both observe and be observed. The extra restriction I imposed pertains only to what constitutes a proper physical system to begin with, not to which of them can constitute an observer. The fact that mind-at-large and all alters are minded does not privilege them over anything else, for according to the proposed framework there is no proper physical system that is *not* minded.

- *The relational interpretation asserts that quantum mechanics provides a complete and self-consistent scheme of description of the physical world.* Indeed, since the physical world of an observer consists in the compound state of the Markov Blanket associated with that observer, quantum mechanics does provide a complete scheme of description of that world. I did suggest earlier that, if we could write down the wave function of a human psyche and formalize the dynamics of its interaction with ψ, we could solve the measurement problem. But this only means that, in practice, we have not yet exhausted the potential of the scheme of description provided by quantum mechanics. My suggestion does not require hidden variables.

The idealist ontology proposed is thus consistent with the relational interpretation and provides an ontological framework for its tenets.

6.13 Solving the qualms of the relational interpretation

We now return to where we started: the philosophical qualms raised by the relational interpretation. The goal of this final section is to show that the ontological framework proposed in Sections 6.8 to 6.11 solves those qualms. Point by point:

- *The intuition of a shared world:* the framework illustrated in Figure 6.4 shows that, even though we do not inhabit the same physical world, we do share a common non-physical environment—namely, mind-at-large. We are all alters of one mind, surrounded like islands by the ocean of its thoughts (ψ). Although each observer lives in its own physical world, this world is created by an interaction—perhaps an interference pattern—between ψ and the observer's own internal state r. Therefore, insofar as the internal state r is similar across observers—a reflection of our common humanity or even of the basic characteristics of life that we share with all organisms—such interaction should, at least in principle, lead to similar worlds.

- *The ontological ground of information:* according to the proposed framework, mind is the sole ontological primitive and ground of all reality. Information is thus given by the discernible qualities of experience, which are themselves patterns of excitation of mind. The problems of (a) why we cannot mentally influence the laws of physics and (b) why we cannot directly access each other's thoughts are both solved by positing dissociation to be a primary natural phenomenon.

- *Relationships without absolutes:* there are no such things. According to the proposed framework, all physical quantities are relationships between *mental* absolutes. A physical quantity encompassed by the sensory state s of an observer consists of a relationship between ψ and the observer's internal state r (see Figure 6.3 again). So there are absolutes: ψ and r. It is just that, in accord with the relational interpretation, these absolutes are not physical quantities.

- *The meaning of 'physical world':* according to the proposed framework, the physical world corresponds to the compound state $s\text{-}a$ of the respective observer's Markov Blanket.

- *The meaning of 'physical system':* according to the proposed framework, only mind-at-large and alters are physical systems. Everything else is just segments of these systems' extrinsic appearances, delineated arbitrarily like figures traced on tree bark.

6.14 Conclusions

I have proposed an idealist framework as ontological underpinning for the relational interpretation of quantum mechanics. According to this framework, a universal mind is the sole ontological primitive underlying all reality. Physical systems consist of dissociated segments of this universal mind, which can observe and be observed by each other. The dissociated segments comprise alters immersed in mind-at-large. Alters have internal states r, which are quantum superposition states. Mind-at-large has state ψ, which is also a quantum superposition state. Alters interact with mind-at-large through mental impingement across their respective dissociative boundaries. This interaction is a quantum observation that creates the physical world of the alter and causes ψ to become correlated with the alter's state. This way, ψ accrues (Shannon) information about all alters. By arising from interactions with ψ, the physical world of each alter can thus reflect the presence and actions of all other alters. I have referred to these reflections as the extrinsic appearances of other alters. Living bodies are the extrinsic appearances, in our respective physical worlds, of other alters.

The proposed ontological framework solves the philosophical qualms raised by the relational interpretation, such as: the intuition that we all share the same external environment, the ontological ground of (Shannon) information, the meaning of physical relationships in the absence of physical absolutes, the nature of the physical world and the criteria for decomposing the world into distinct physical systems.

It is hoped that the combination of the relational interpretation with the idealist framework articulated in this paper offers a promising avenue to make sense of reality in a parsimonious manner, consistent with experimentally confirmed contextuality.

Part III

Refuting objections

The introduction of 'something else' violates our mode of thought and the convenience of its habitual operations. A 'something else' disturbs minds that mistake comfortable thinking with clarity of thought.
James Hillman: *The Soul's Code.*

There may be times when what is most needed is ... a different 'slant'; I mean a comparatively slight readjustment in our *way* of looking at the things and ideas on which attention is already fixed.
Owen Barfield: *Saving the Appearances.*

Chapter 7

Preamble to Part III

For centuries, Western society has lived under the ontologies of dualism and, more recently, physicalism. Both of these ontologies entail that there is an objective reality outside and independent of mind; that is, that the world is essentially non-phenomenal. Because the denial of this notion is precisely the key tenet of idealism, the ontology laid out in the previous two chapters goes against centuries of cultural indoctrination. It is thus not surprising that, when confronted with an argument for idealism, many people are unable to hear the argument for what it is, but project onto it, instead, historical biases and unexamined assumptions. In other words, they do not hear what the idealist actually says, but what they instead expect the idealist to be saying. They remain largely unable to think within the internal logic and terminology put forward by the idealist, constantly injecting their own logical bridges and expectations into the idealist's reasoning. The result is that they succeed in burning down a straw man, which—ironically—only reinforces their own misapprehensions.

Because of the formidable cultural momentum behind the notion of an objective physical world distinct from mind, most educated people today can promptly leverage a ready-made, culturally-sanctioned list of objections against idealism, which they then hurl at the idealist with righteous condescendence. This is understandable, for we are all overwhelmed with physicalist (or dualist) assumptions since a very tender age; not only through formal education, but also through the very structure of the language we are taught (for instance, the separation between subject and object, as well as the distinction between verb and noun, play right into physicalist and dualist intuitions).

The idealist must accept that the knee-jerk resistance to his or her views is largely inevitable, given present cultural circumstances. It would be naïve to expect of the average educated person the level of cultural detachment necessary to give idealism a truly impartial hearing. We are all immersed in myriad unexamined cultural assumptions and conventional patterns of thought. It is the unavoidable task of the idealist to patiently identify and expose these hidden assumptions and faulty thought patterns, one by one, whilst persisting in repeated elucidations of his or her argument. Only in this manner can the logic behind the idealist's argument eventually pierce through the cultural shield.

In this spirit, Chapter 8 attempts to address the many objections often leveled against idealism. Although three of these objections have already been briefly addressed earlier in Chapter 5, they are more extensively refuted in Chapter 8. In addition, the sequence of refutations is deliberately arranged so that the reader does not need to be acquainted with Part II to follow the argument. As such, Chapter 8 comprises, in and of itself, a rather more generic articulation of the basic idealist ontology proposed in this book.

One particular objection is special because it isn't merely a cultural artifact, but indicates some legitimate difficulties: a necessary implication of the ontology proposed in Part II is that a person's metabolism—*all of it*—is the extrinsic appearance of the person's *conscious* inner life. This is reasonable enough for certain patterns of brain activity known to correlate with experiences accessible through introspection, but what about metabolism beyond the brain, such as e.g. liver and kidney function? And what about the metabolic activity taking place in, say, the person's left big toe? If the ontology proposed in Part II is correct, then liver, kidney and—yes—toe function should all be the extrinsic appearance of *conscious experiences* as well. Yet, try as we might, these experiences do not seem to be accessible through introspection.

Moreover, even if we were to look at the brain alone, ignoring the metabolism in the rest of the body, recent studies in neuroscience and psychology suggest the presence of *un*conscious mental processes in the brain. This, if true, would already contradict the idealist tenet that all reality is *in consciousness*.

Chapter 9 bites these bullets and argues that, despite appearances to the contrary, we have no clear reason to believe that any mental process is *truly* unconscious. Instead, I hope to show that there are, in fact, very good reasons to think that what we regard as unconscious mental processes correspond to an *illusion* of unconsciousness, which results from dissociative states or lack of metacognition. And once these two mechanisms—dissociative states and lack of metacognition—are identified, they can explain why conscious experiences corresponding to areas of the living body beyond the nervous system can't be accessed through introspection.

There is one more reason why the argument laid out in Chapter 9 is important at the present juncture. Because there is great societal pressure on neuroscience today to unveil the biological foundations of consciousness—which, of course, places neuroscience in an impossible position—new studies are often portrayed as if they represented progress in this direction. Many of these studies, however, tackle merely the mechanisms behind the rise of *metacognition*, not of consciousness itself. Metacognition happens when we realize *that* we are having an experience, but the experience itself could be going on since long before the rise of metacognition. A recent study, for instance, has identified neural activity that accompanies an 'Aha!' moment (Kang et al. 2017), which is a textbook example of metacognition: the moment when we realize *that* we know something. Yet, it has been claimed that "this study offers new hope that the biological foundations of consciousness may well be within our grasp" (Zuckerman Institute 2017). This is a *non sequitur*, for what is arising is metacognition, not necessarily consciousness; the latter could have been there all along (Chapter 9 elaborates on all this in detail). As such, this and other related studies say nothing about how the *qualities* of an experience arise; they simply identify neural correlates of the moment when we know *that* we have the experience. In addressing metacognition, these studies *presuppose* consciousness, as opposed to reducing it.

You see, where physicalism fails is in explaining how physical parameters give rise to the qualities of experience, this being the essence of the "hard problem of consciousness" (Chalmers 2003). Explaining how consciousness—once it is there—can turn in upon itself in order to experience knowledge *of* its experiences is *not* the hard problem, but a comparatively 'easy problem.' By conflating metacognition with consciousness proper, and then making progress in identifying the neural correlates of the rise of metacognition, neuroscience replaces the 'hard problem' with an 'easy problem' and then pretends to be explaining how consciousness arises. While refuting a reasonable objection to idealism, Chapter 9 exposes this *un*reasonable sleight of hand, which has so far prevented society from confronting the fact that neuroscience has made precisely *zero* progress in explaining consciousness under physicalist assumptions.

All in all, Part III tackles all the objections to idealism I could think of. In the interest of completeness, I have even added one or two objections that most philosophers wouldn't consider serious enough to merit a refutation. My hope is that no remotely reasonable objection will be left unaddressed.

Chapter 8

On the plausibility of idealism: Refuting criticisms

This article first appeared in *Disputatio: International Journal of Philosophy*, ISSN: 0873-626X, Vol. 9, No. 44, pp. 13-34, in May 2017. Founded in 1996 and reaching in 2016 the second highest ranking (Q2) in the Scimago Journal Rankings — a well-recognized measure of an academic journal's prestige — *Disputatio* is published by the Philosophy Center of the University of Lisbon.

8.1 Abstract

Several alternatives vie today for recognition as the most plausible ontology, from physicalism to panpsychism. By and large, these ontologies entail that physical structures circumscribe consciousness by bearing phenomenal properties within their physical boundaries. The ontology of idealism, on the other hand, entails that all physical structures are *circumscribed by* consciousness in that they exist *solely as* phenomenality in the first place. Unlike the other alternatives, however, idealism is often considered implausible today, particularly by analytic philosophers. A reason for this is the strong intuition that an objective world transcending phenomenality is a self-evident fact. Other arguments — such as the dependency of phenomenal experience on brain function, the evidence for the existence of the universe before the origin of conscious life, etc. — are also often cited. In this essay, I will argue that these objections against the plausibility of idealism are false. As such, this essay seeks to show that idealism is an entirely plausible ontology.

8.2 Introduction

The mainstream physicalist ontology posits that reality is constituted by irreducible physical entities—which Strawson (2006: 9) has called "ultimates"—outside and independent of phenomenality. According to physicalism, these ultimates, in and of themselves, do not instantiate phenomenal properties. In other words, there is nothing it is like to be an ultimate, phenomenality somehow emerging only at the level of complex arrangements of ultimates. As such, under physicalism phenomenality is not fundamental, but instead reducible to physical parameters of arrangements of ultimates.

What I will call 'microexperientialism,' in turn, posits that there is already something it is like to be at least some ultimates (Strawson et al. 2006: 24-29), combinations of these experiencing ultimates somehow leading to *more complex* experience. As such, under microexperientialism phenomenality is seen as an irreducible aspect of at least some ultimates. The ontology of *pan*experientialism (Griffin 1998: 77-116, Rosenberg 2004: 91-103, Skrbina 2007: 21-22) is analogous to microexperientialism, except in that the former entails the stronger claim that *all* ultimates instantiate phenomenal properties.

Micropsychism (Strawson et al. 2006: 24-29) and panpsychism (Skrbina 2007: 15-22) are analogous—maybe even identical—to microexperientialism and panexperientialism, respectively, except perhaps in that some formulations of the former admit cognition—a more complex form of phenomenality—already at the level of ultimates, as an irreducible aspect of these ultimates.

While microexperientialism, panexperientialism, micropsychism and panpsychism entail that bottom-up combinations of simple subjects give rise to more complex ones, such as human beings, cosmopsychism (Nagasawa & Wager 2016) takes the opposite route. Indeed, "the first postulate of cosmopsychism is that *the cosmos as a whole is the only ontological ultimate* there is, and that it is *conscious*" (Shani 2015: 408, original emphasis).

Finally, the ontology of idealism is characterized by a combination of two propositions: (a) phenomenal consciousness is irreducible; *and* (b) everything else—the whole of nature—is reducible to a unitary and universal phenomenal consciousness (henceforth, I shall refer to phenomenal consciousness simply as 'consciousness').

Idealism may be consistent with—even identical to—certain interpretations of cosmopsychism. According to Shani, for instance, cosmopsychism entails that "an omnipresent cosmic consciousness is the single ontological ultimate there is" (2015: 390). This perfectly embodies the defining tenet of idealism insofar as it implies that everything—*including the physical*—can be reduced to the phenomenal. Shani also writes that matter is the cosmos "in its *appearance* as exterior complement to the *subjective* realities of created selves" (2015: 412, emphasis added). The notion that matter is the phenomenal appearance of equally phenomenal dynamics is also eminently idealist. Therefore, these interpretations of cosmopsychism are essentially indistinguishable from idealism and I shall, henceforth, refer to them simply as idealism.

Other possible interpretations of cosmopsychism entail that the cosmos as a whole bears phenomenal properties—that is, has inner life—but also has an aspect—the physical universe we can measure—that is irreducible to these phenomenal properties. Naturally, this implies a form of dual-aspect monism, *a la* Spinoza (Skrbina 2007: 88). Indeed, under these views the cosmos can still be said to be *conscious*, but not *in consciousness*. In the former case, the cosmos *bears* phenomenality; in the latter—which is the idealist view—the cosmos is *constituted by* phenomenality. Interpretations of cosmopsychism that are not consistent with idealism shall not be further addressed in this paper.

In what follows, I will attempt to rebut the most common objections to the plausibility of idealism. I will seek to show that these objections are based on circular reasoning, conflation, unexamined assumptions, and several other misconceptions.

8.3 The felt concreteness objection

English poet Samuel Johnson is said to have argued against Bishop Berkeley's idealism by kicking a large stone while exclaiming: "I refute it *thus!*" (Boswell 1820: 218). Johnson was clearly appealing to the felt concreteness of the stone to suggest that it could not be just a figment of imagination. Indeed, the felt concreteness of the world is probably the main reason why people intuitively reject the notion that reality unfolds in consciousness. If a truck hits you, you will hurt, even if you are an idealist.

However, notice that appeals to concreteness, solidity, palpability and any other quality that we have come to associate with things outside consciousness are still appeals to phenomenality. After all, concreteness, solidity and palpability are *qualities of experience*. What else? A stone allegedly outside consciousness, in and by itself, is entirely abstract and has no qualities. If anything, by pointing to the *felt* concreteness of the stone Johnson was implicitly suggesting the primacy of experience over abstraction, which is eminently idealist.

We have come to automatically *interpret* the felt concreteness of the world as evidence that the world is outside consciousness. But this is an unexamined artifact of subliminal thought-models. Our only access to the world is through sense perception, which is itself phenomenal. The notion that there is a world outside and independent of the phenomenal is an *explanatory model*, not an empirical fact. No phenomenal quality can be construed as direct evidence for something outside phenomenality.

8.4 The private minds objection

As discussed in the Introduction, under idealism there is only one universal consciousness. Yet, at a personal level, our mental lives are clearly separate from one another. I do not have direct access to your thoughts and feelings and, presumably, neither do you to mine. Moreover, I do not seem to be aware of what is happening across the galaxy and, presumably, neither are you. So, if all reality is reducible to one universal consciousness, how can there be separate private minds such as yours and mine?

To make sense of this under idealism, we need to review a mental condition called *dissociation* (Braude 1995, Kelly et al. 2009: 167-174 & 348-352, Schlumpf et al. 2014, Strasburger & Waldvogel 2015). Indeed, it is now well established in psychiatry that mental contents can undergo "a disruption of and/or discontinuity in [their] normal integration" (Black & Grant 2014: 191). This normal integration of mental contents takes place through chains of cognitive associations: a perception may evoke an abstract idea, which may trigger a memory, which may inspire a thought, etc. These associations are *logical*, in the sense that e.g. the memory inspires the thought because of a certain *implicit logic* linking the two. Integrated mentation can thus be modeled, for ease of visualization, as a connected directed graph. See Figure 8.1a. Each vertex in the graph represents a particular mental content and each edge a cognitive association logically linking mental contents together. Every mental content in the graph of Figure 8.1a can be reached from any other mental content through a chain of cognitive associations. Dissociation, in turn, can be visualized as what happens when the graph becomes disconnected, such as shown in Figure 8.1b. Some mental contents can then no longer be reached from others. Following the psychiatric convention, I shall refer to the subgraph with grey vertices as a (dissociated) *alter*.

Because cognitive associations are essentially logical, as opposed to spatio-temporal, the scheme of representation in Figure 8.1 allows for the *simultaneous* experience of multiple mental contents linked together in a connected subgraph. This is empirically justifiable: a perception, for instance, can be experienced *at the same time* as the thoughts it evokes and the

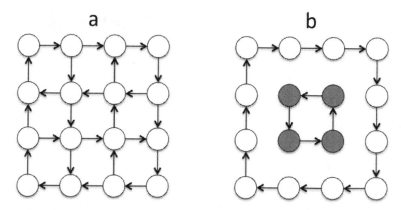

Figure 8.1: A connected graph (a) illustrating normal integration of mental contents, and a disconnected graph (b) illustrating dissociation and the corresponding formation of an alter (subgraph in grey).

emotions evoked by these thoughts. Moreover—and by the same token—the two disconnected subgraphs in Figure 8.1b can also represent two *concurrently conscious* subjects of experience. The substantiation for this is again empirical: there is compelling evidence that different alters of the same psyche can be co-conscious (Kelly et al. 2009: 317-322, Braude 1995: 67-68).

An alter loses direct access to mental contents surrounding it, *but remains integral to the underlying consciousness that constitutes it*. The disconnection between an alter and surrounding mental contents is logical, not ontic. As an analogy, a database may contain entries that are not indexed and, therefore, cannot be reached, but this does not physically separate those entries from the rest of the database.

Dissociation can coherently explain how seemingly separate but concurrently conscious subjects of experience—such as you and me—can form under idealism: each is an alter of universal consciousness. And because each alter becomes unable to evoke the mental contents of another, their respective inner lives acquire a seemingly private character, even though they remain integral to the underlying consciousness that constitutes them.

8.5 The stand-alone world objection

If all there is is consciousness, does the world continue to exist when not consciously observed by a living being? A negative answer to this question seems extremely implausible yet difficult to avoid under idealism. Bishop Berkeley has famously attempted to circumvent it by appealing to a divinity, as captured in Ronald Knox's limerick, *God in the Quad*:

> There was a young man who said 'God
> Must find it exceedingly odd
> To think that the tree
> Should continue to be
> When there's no one about in the quad.'
> Reply:
> 'Dear Sir: Your astonishment's odd;
> I am always about in the quad.
> And that's why the tree
> Will continue to be
> Since observed by, Yours faithfully, God.'

Legitimate as an appeal to a divinity might have been in Berkeley's time, today more rigor is expected from a viable ontology. So how do we solve the problem of a stand-alone world under idealism?

With reference to the discussion in the preceding section, notice that, by definition, mental contents inside an alter of universal consciousness cannot directly evoke mental contents outside the alter, or vice-versa. But they can still *influence* or *impinge on* each other. Indeed, mental impingement across a dissociative boundary is empirically known. Lynch and Kilmartin (2013: 100), for instance, report that dissociated feelings can dramatically affect our thoughts, while Eagleman (2011: 20-54) shows that dissociated expectations routinely mold our perceptions. We can visualize this as in Figure 8.2a, wherein the partial overlap of adjacent vertices internal and external to the alter (cf. Figure 8.1b) represents mental impingement across its dissociative boundary.

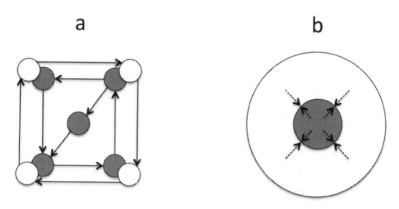

a **b**

Figure 8.2: Mental contents impinging on the dissociative boundary of an alter, illustrated in two different ways (a) and (b).

Figure 8.2b illustrates the exact same thing according to a simplified representation: the broader consciousness is represented as a white circle with an alter represented as a grey circle within it. The dashed arrows represent the impingement of external and internal mental contents on each other, across the alter's boundary. I will henceforth use this simplified representation.

Now notice that mental contents of universal consciousness that surround—*but remain external to*—an alter can impinge on the alter's boundary from the outside. Under idealism, it can be coherently argued that this is what gives rise to sense perceptions: the physical world around us is the extrinsic *appearance* on the screen of perception of phenomenality surrounding our respective alter. See Figure 8.3.

The stand-alone character of the world can thus be coherently explained: the world is a perceptual representation of phenomenality dissociated from our personal psyche and,

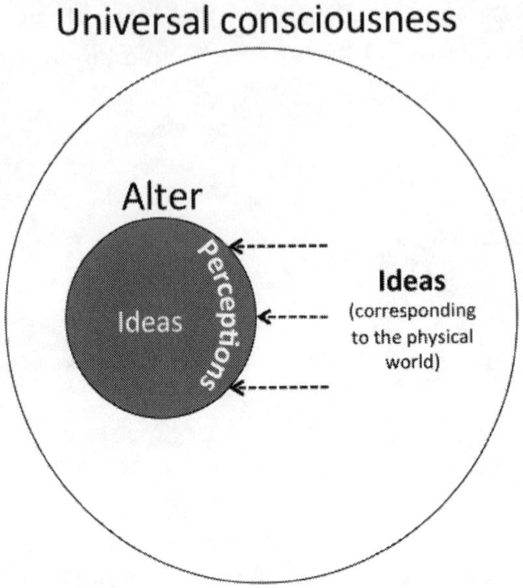

Universal consciousness

Figure 8.3: Mental contents of universal consciousness surrounding an alter can cause the alter's sense perceptions by impinging on its dissociative boundary.

as such, independent of our personal inner life. That which underlies the physical world we perceive continues to exist—in the form of phenomenality outside our respective alter—even as we sleep.

8.6 The autonomy of nature objection

A closely related objection is this: nature unfolds according to patterns and regularities—the 'laws of nature'—independent of our personal volition. Human beings cannot change these laws. But if nature is in consciousness, should that not be possible by a mere act of imagination?

This objection can be rebutted along the same lines as the previous one. However, there is a more direct and intuitive refutation. Notice that the implicit assumption here is that all mental activity is acquiescent to volition, which is patently false even in our own personal psyche. After all, by and large we cannot control our dreams, nightmares, emotions, and even many of our thoughts. They come, develop and go on their own terms. At a pathological level, schizophrenics cannot control their visions and people suffering from obsessive-compulsive disorder are constantly at the mercy of oppressive thoughts. There are numerous examples of conscious activity that escapes the control of volition. Often, we do not even recognize this activity as our own; that is, we do not identify with it. It unfolds as autonomous, seemingly external phenomena, such as dreams and schizophrenic hallucinations. Yet, all this activity is unquestionably within consciousness. We perceive it as separate from ourselves because the segment of our psyche that gives rise to this activity is dissociated from the ego, the segment with which we do identify.

So that there is activity in universal consciousness that we do not identify with and cannot control is entirely consistent with idealism. This activity is simply dissociated from our ego and its sense of volition.

8.7 The shared world objection

If all reality is in consciousness, then the world is akin to a dream. As such, idealism implies that we are all partaking in roughly the same dream. Yet, since our bodies are separate, we cannot be sharing a dream; or so the objection goes.

The objection begs the question by implicitly assuming that the body circumscribes dreaming consciousness, as opposed to the other way around. Only under this assumption does the impossibility of sharing a dream follow from the fact that bodies are separate. But under idealism, it is the body that is in universal consciousness, not consciousness in the body. Once this is properly understood according to the framework developed in the preceding sections, the rebuttal of this objection becomes rather straightforward: we all seem to inhabit the same world because our respective alters are surrounded by the same universal field of phenomenality, like whirlpools in a single stream. See Figure 8.4, which simply extends Figure 8.3 to multiple alters.

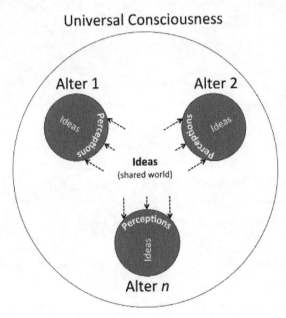

Figure 8.4: Alters of universal consciousness and their shared world.

8.8 The natural order objection

The world we perceive around ourselves is governed by stable and orderly natural laws. Therefore, if the contents of perception are a representation of phenomenality in universal consciousness, then this phenomenality must be stable and orderly at root. But our own personal thoughts and emotions are notoriously unstable and disorderly. So how plausible is it that the order and stability we discern in the laws of nature represent thoughts or emotions in universal consciousness?

The misconception here, of course, is that of anthropomorphization: to attribute to universal consciousness as a whole cognitive characteristics known only in small dissociated segments of it, such as human beings. Nothing in idealism precludes the possibility that phenomenality in universal consciousness unfolds according to very stable and orderly patterns and regularities, whose extrinsic appearance corresponds to the laws of nature. That our human thoughts and emotions seem rather reactive and unstable is a product of evolution under the pressures of natural selection within a particular planetary ecosystem. At a universal level, consciousness has not undergone such evolutionary pressures.

Under physicalism, the laws of nature are seen as irreducible causal patterns *somehow* built into the fabric of the cosmos. It is the dynamic unfolding of these patterns that leads to the order and complexity we see around ourselves. Under idealism, such irreducible causal patterns are posited to be *somehow* built into universal consciousness itself, instead of an objective fabric of space-time. Yet, beyond this distinction, *they are the same patterns* that physicalism entails, as inherent to consciousness as physical laws are allegedly inherent to the fabric of space-time. Idealism poses no extra difficulty than physicalism in this regard.

This can be better understood with a simple terminology move. Certain schools of psychology speak of "psychological archetypes": innate, built-in templates according to which mental dynamics unfold (Jung 1991). As such, we can say that, under idealism, the laws of nature are the *archetypes of universal consciousness*. They are built-in templates according to which the 'vibrations' of universal consciousness—that is, phenomenality—develop, analogously to how the physical constraints of a vibrating surface determine its natural modes of vibration.

8.9 The equivalence objection

As we have seen in Sections 8.5 to 8.7, idealism acknowledges that there is a world outside *personal* psyches, since personal psyches are but dissociated segments of a broader universal consciousness. The objection, then, is that the notion of a broad stream of phenomenality outside personal psyches is equivalent to the physicalist postulate of a world outside consciousness.

Except for solipsism, any viable ontology must entail at least one inference beyond direct experience. This is necessary to make sense of the fact that we all inhabit the same world beyond ourselves and are unable to change its governing laws. For this reason, physicalism infers the existence of a universe outside consciousness, which we all inhabit. Idealism, on the other hand, infers simply that consciousness itself extends beyond its face-value personal boundaries. This way, while physicalism postulates a fundamentally new ontological class next to experience, idealism simply *extrapolates* the boundaries of consciousness—the sole undeniable ontological class and primary datum of existence—beyond those we can probe directly. To put it metaphorically, while idealism makes sense of reality by inferring that the Earth extends beyond the visible horizon, physicalism does so by inferring the existence of an isomorphic[1] but ontologically distinct 'shadow' Earth. Clearly, the former is a more parsimonious inference and, as such, not equivalent to the latter.

More importantly, the implications of idealism are radically different from those of physicalism. For instance, while physicalism implies that consciousness ends upon the death of the body, idealism implies merely the end of the corresponding *dissociation*, not of consciousness proper. I have elaborated on further differences in implications elsewhere (Kastrup 2015: 185-198).

8.10 The primacy of brain function objection
Not only are there (a) clear correlations between specific patterns of brain activity and reported inner experience (Koch 2004), we

---

1 Isomorphism entails a correspondence of form.

know that (b) physical interference with the brain—such as head trauma and the use of psychoactive drugs—can influence one's inner life rather dramatically. This may seem to suggest an arrow of causation pointing from a physical body outside consciousness to phenomenality, which would contradict idealism.

To make sense of observation (a), we need to briefly recapitulate earlier discussions. As we have seen in Section 8.4, under idealism private minds—such as our own human psyche—can be explained as dissociated alters of universal consciousness. We have also seen in Section 8.5 that the stand-alone world around us can be explained as the extrinsic appearance of phenomenality surrounding—but outside—our respective alter. Now, from the point of view of a given alter *A*, nothing prevents the dissociated mental activity of an alter *B* from being part of the phenomenality surrounding *A*. *B* is then part of *A*'s world and, as such, must also have an extrinsic appearance on *A*'s screen of perception. In other words, there must be something alters *look like* from a second-person point of view. And since we know from direct experience that our private inner life extends only to the boundaries of our metabolizing body—after all, we cannot perceive things that do not impinge on our skin or other sense organs, or move anything beyond our own body through direct intention—metabolizing bodies seem *prima facie* to be the extrinsic appearance of dissociated alters of universal consciousness. If so, this means that all living beings have private inner lives in some way analogous to our own, but tables and chairs do not. The latter are simply aspects of the *inanimate* universe, which, *as a whole*, is the extrinsic appearance of phenomenality *outside all alters*.

Brain activity, of course, is integral to a metabolizing human body. Therefore, under idealism, brain activity is simply part of what one's private inner experiences—self-reflective and otherwise, as I will elaborate upon in the next section—*look like* from across a dissociative boundary. To put it another way, one's brain activity is part of a phenomenal representation of one's inner life. And of course, a representation *must* correlate with the phenomenal process it is the appearance of, without requiring anything ontologically distinct from consciousness. That this correlation is empirically observed is thus entirely consistent with idealism.

A possible counterargument here is this: the patterns of neural activity one can measure with functional brain scanners can be enormously complex in terms of information content; perhaps more complex than the contents of consciousness we have introspective access to. What does the extra complexity then correspond to? The key to answering this question is in the next section, wherein a distinction will be made between contents of consciousness we have introspective access to—that is, can self-reflect upon—and contents of consciousness that, despite still being experienced, fall outside the reach of introspection. The extra complexity, insofar as it indeed is the case, corresponds to the latter.

Regarding observation (b) of the objection, the suggested arrow of causation is based on an unexamined but pervasive assumption: that the physical is in some sense distinct from, yet causally effective upon, the phenomenal. This is precisely what idealism denies. Under idealism, the physical is simply the contents of perception, a particular type of phenomenality. As such, what we call 'physical interference with the brain' is the extrinsic appearance of phenomenality external to an alter that disrupts the inner experiences of the alter from across its dissociative boundary. The disruption 'pierces through' the boundary, so to speak. And that certain types of phenomenality disrupt other types of phenomenality is not only entailed by idealism, but also empirically trivial. After all, our thoughts disrupt our emotions—and vice-versa—every day. For the same reason that thoughts disrupt emotions, 'physical interference with the brain' disrupts an organism's inner life. None of this contradicts idealism.

8.11 The unconscious mentation objection

In Libet's now famous experiments (1985), neuroscientists were able to record, a fraction of a second *before* subjects reported making a decision to act, mounting brain activity associated with the initiation of a simple voluntary action. At first sight, this would seem to indicate that decisions are made in a neural substrate outside consciousness, thereby contradicting idealism. I use Libet's experiments here merely as an example, for today we know of many other instances of seemingly unconscious mentation, such as moving one's foot halfway to the brake pedal before one becomes aware of danger ahead (Eagleman 2011: 5). Under idealism, since everything is in consciousness, there cannot be such a thing as unconscious mentation. So what is going on?

The misconception here is a conflation of consciousness proper with a particular *configuration of* consciousness. Indeed, to report an experience—such as making a decision to act or seeing danger ahead—to another or to oneself, one has to *both* (a) have the experience *and* (b) know *that* one has the experience, which Schooler (2002) called a "re-representation." In other words, one can only report phenomenality that one is self-reflectively aware of at a metacognitive level. But self-reflection is just a particular configuration of consciousness, whereby consciousness turns in upon itself to experience knowledge of its own phenomenality (Kastrup 2014: 104-110). Nothing precludes the possibility that phenomenality takes place outside the field of self-reflection. In this case, we cannot report the phenomenality—not even to ourselves—because we do not know *that* we experience it.

The argument above is not idiosyncratic, for the existence of unreportable phenomenality is well established in neuroscience today (Tsuchiya et al. 2015, Vandenbroucke et al. 2014). Indeed, as elaborated upon by Schooler (2002), reportability is an *extra* function at a metacognitive level, on top of phenomenality proper. So the possibility that presents itself to us is that *all* mentation is actually conscious, even though we cannot report much of it. As such, the decisions made by Libet's subjects could well have been made in consciousness, but outside the field of self-reflection. The corresponding phenomenality then entered this field a fraction of a second later, thereby becoming reportable. Analogously, drivers may consciously see danger ahead before they can tell themselves *that* they see danger ahead. The *appearance* of unconscious mentation due to unreportability does not contradict idealism.

8.12 The unconsciousness objection

Along similar lines, the idea here is that, when we e.g. faint or undergo general anesthesia, we become seemingly unconscious. Yet, we do not cease to exist because of it, which may seem to contradict the idealist tenet that our body is the extrinsic appearance of conscious inner life.

Let us consider this more carefully. Imagine that you wake up in the morning after hours of deep sleep. You may remember nothing of what happened during those preceding hours, concluding that you were unconscious all night. Then, later in the day, you suddenly remember that you actually had a very intense dream. So you were not unconscious all night, you simply could not *remember* your experiences.

Indeed, all we can assert with confidence upon coming round from episodes of *seeming* unconsciousness is that we cannot *remember* phenomenality occurring during those episodes. The actual *absence* of phenomenality is impossible to assert with confidence. As a matter of fact, many things we have traditionally associated with unconsciousness are now known to entail intense experiences. For instance, fainting caused by e.g. asphyxiation, strangulation or hyperventilation is known to correlate with euphoria, insights and visions (Neal 2008: 310-315, Rhinewine & Williams 2007, Retz 2007). G-force-induced loss of consciousness (G-LOC) is also known to correlate with "memorable dreams" (Whinnery & Whinnery 1990). There is even evidence for "implicit perception" during general anesthesia (Kihlstrom & Cork 2007).

Sleep, of course, is known to correlate with dreams. But even during phases of sleep wherein electroencephalogram readings show no dream-related neural activity, there are other types of activity that may correlate with non-recallable phenomenality distinct from dreams. Indeed, this is precisely what a recent study points out: "there are good empirical and theoretical reasons for saying that a range of different types of sleep experience, some of which are distinct from dreaming, can occur *in all stages of sleep*" (Windt, Nielsen & Thompson 2016: 871, emphasis added). The authors identify three different categories of sleep experiences distinct from dreams: (a) non-immersive imagery and sleep thinking, (b) perceptions and bodily sensations, and (c) 'selfless' states and contentless sleep experiences that may be similar to those reported by experienced meditators.

As such, what the empirical data shows is that episodes of seeming unconsciousness are associated with an impairment of memory access, but not necessarily with absence of phenomenality. As a matter of fact, there are strong indications, as mentioned above, that the opposite is true.

8.13 The solipsism objection

Some conflate idealism with solipsism, the notion that the world is one's *personal* dream, all other living creatures being just figments of one's personal imagination. Under solipsism, there is nothing it is like to be other people; they have no inner life; they exist only as appearances in the personal psyche of the dreamer. As such, whatever empirical evidence one brings to bear and whatever one says to a solipsist must be regarded by the solipsist as figments of his or her own imagination, which renders solipsism unfalsifiable. So the objection here is that, by being unfalsifiable, solipsism—and therefore idealism—is beneath philosophical debate.

Naturally, idealism is not solipsism. Under idealism, there *is* something it is like to be other living creatures; they also have private inner lives. So idealists take other people seriously as legitimate sources of reported experiences and views, not just as figments of one's own imagination. Moreover, idealists acknowledge that *there is a world outside and independent of their personal (dissociated) psyche*, as discussed in Sections 8.5 to 8.7. They simply do not acknowledge that this world is ontologically distinct from consciousness itself. Indeed, by acknowledging that dissociation in universal consciousness implies a world outside their own personal mentation, idealists look upon this world in a way entirely compatible with naturalism and scientific inquiry.

Unlike solipsism, idealism has the burden to explain observations non-trivially. Consider three basic facts that are often used to justify physicalism: (a) the laws of nature are independent of our personal volition; (b) we all seem to inhabit the same world; and (c) there are tight correlations between observable brain activity and reported inner life. Solipsism trivializes all three facts *in lieu* of actually making sense of them: the solipsist allegedly dreams them all up, rather arbitrarily. The idealist, on the other hand, by acknowledging the inner lives of other people and the autonomous nature of the world, has the burden to reconcile these three facts with the notion that reality unfolds in consciousness. If idealism is correct, (a) how come we cannot simply imagine a different and better world? If the world is akin to a dream in consciousness, (b) how come we are all having the same dream? If consciousness is not generated by the brain, (c) how come there are such tight correlations between brain activity and inner experience? These questions have already been answered in Sections 8.6, 8.7 and 8.10, respectively. The important point here is this: idealism is falsifiable in that, if it cannot answer these and other questions in terms of universal consciousness alone, it must be discarded.

8.14 The cosmological history objection

There is overwhelming evidence for the existence of the universe before conscious life arose. Therefore—or so the objection goes— it is untenable to say that the universe exists in consciousness. This may strike some readers as obviously question-begging— which, of course, it is—but please bear with me for the sake of completeness.

The implicit assumption here is that consciousness arises only with biology, as a product of biology. Naturally, this is precisely what idealism denies. Under idealism, biology is merely the extrinsic appearance of *dissociated, local differentiations of consciousness* (that is, alters), not the constituent or generator of consciousness. There was universal consciousness before such dissociated, local differentiations arose. And there was phenomenality in this universal consciousness corresponding to the inanimate universe prior to the origin of life.

8.15 The implausibility of cosmic inner life objection

The last objection I will address in this essay is, like the first, purely intuitive. It asks rhetorically: How plausible is it that the inanimate universe as a whole is the extrinsic appearance of some kind of universal inner life? The intuitive appeal of the question is understandable. After all, we only have introspective access to our own (dissociated) personal inner life, so to gauge the presence of other or broader inner life we depend on perceivable external indicators. In other people and animals, these indicators are their behavior. But within the extremely small range of space and time in which we live our lives—and even in which human history as a whole has unfolded—we simply cannot perceive any intuitively-appealing indicator of universal inner life.

Yet, we can approach the question from a different angle. Consider a living brain exposed by surgeons during an operation. It is a very concrete object that can be seen, touched, cut, cauterized, etc. It is composed of the same types of atoms and force fields that make up the universe as a whole. There is nothing magical about a brain insofar as we can gauge on the screen of perception. And neither can we discern any intuitively-appealing indicator of inner life by simply looking at an exposed brain.

Nonetheless, we all know that 'behind' the living brain lies the entire inner life of a person, with love affairs and heartbreaks, successes and disappointments, great adventures and quiet introspective insights, great joy and indescribable suffering. 'Behind' that very concrete object under the surgeon's scalpel there lies a world of phenomenality. Counterintuitive or not, this is the way nature is: what we call physical structures—such as living brains—*can* correspond in some way to rich phenomenality. We may not know *how* this is so, but we do know *that* it is so.

Therefore, unless we solve the "hard problem of consciousness" (Chalmers 2003) and explain what makes brains different from the inanimate universe as a whole in this regard, if brains correspond to inner life it is not at all implausible that the inanimate universe as a whole could as well. After all, brains are made of the same 'stuff' that the rest of the universe is also made of.

One could argue at this point that only particular structural and functional organizations of this 'stuff,' as found in brains, are conducive to the kind of information processing associated with human inner life. For instance, Tononi (2004) has shown that *reportable* experiences correlate only with complex networks of information integration in the brain. Although it has recently been shown that there are structural similarities between brains and the universe at its largest scales (Krioukov et al. 2012),[2] it is implausible that analogous information integration takes place at a universal level. The distances and signal propagation times involved do not permit it (Siegel 2016).

However, the hypothesis offered here is not that the universe has *human-like* cognition and associated information integration. As a matter of fact, the hypothesis is not even that the universe has *cognition*, defined as the capacity to acquire knowledge or understanding. Instead, the claim is simply that there is *raw experience*—qualia, pure and simple—associated with the universe as a whole, which does not require anything like the kind of information integration underlying human self-reflection.

8.16 Conclusions

Idealism is a unique ontology in that, unlike physicalism and panpsychism, it asserts that physical structures are circumscribed by consciousness, as opposed to the other way around. Yet, analytic philosophy has traditionally considered idealism implausible. In this essay, I have argued that the alleged implausibility of idealism is based on misconceptions, such as:

- Unfounded intuition—e.g. taking the concreteness of the world to indicate its independence from consciousness, or asserting the implausibility of universal inner life;
- Lack of philosophical imagination—e.g. assuming that multiple private minds and a stand-alone world cannot be coherently reduced to a single universal consciousness;
- Demonstrably wrong assumptions—e.g. that all mental activity is acquiescent to volition;
- Question-begging—e.g. arguing that different people

---

2 This conclusion has been confirmed and amplified by a later study done by Franco Vazza and Alberto Feletti (2017).

cannot share a dream because their bodies are separate, and arguing that the universe cannot be in consciousness because it existed before conscious life first arose;

- Anthropomorphization—e.g. taking all conceivable processes in consciousness to necessarily be unstable and disorderly;
- Failure to understand the implications of idealism—e.g. asserting that a field of phenomenality outside personal psyches is equivalent to a physical world outside phenomenality;
- Unexamined assumptions—e.g. that the physical is in some sense distinct from, yet causally effective upon, the phenomenal;
- Conflation—e.g. conflating consciousness proper with self-reflection, conflating unconsciousness with failure to recall phenomenality, and conflating idealism with solipsism.

As such, idealism is an entirely plausible ontology that may offer the most parsimonious and explanatorily powerful option yet to make sense of reality.

Chapter 9

There is an unconscious, but it may well be conscious

This article first appeared in *Europe's Journal of Psychology*, ISSN: 1841-0413, Vol. 13, No. 3, pp. 559-572, on 31 August 2017. *Europe's Journal of Psychology* is published by *PsychOpen*, arguably Europe's main open-access academic publisher in the field of psychology, which is operated by the Leibniz Institute for Psychology Information, Trier, Germany. A summary of this article has also appeared in *Scientific American* on 19 September 2017.[1]

9.1 Abstract

Depth psychology finds empirical validation today in a variety of observations that suggest the presence of causally effective mental processes outside conscious experience. I submit that this is due to misinterpretation of the observations: the *subset* of consciousness called 'meta-consciousness' in the literature is often mistaken for consciousness proper, thereby artificially creating space for an 'unconscious.' The implied hypothesis is that *all* mental processes may in fact be conscious, the *appearance* of unconsciousness arising from our dependence on self-reflective introspection for gauging awareness. After re-interpreting the empirical data according to a philosophically rigorous definition of consciousness, I show that two well-known phenomena corroborate this hypothesis: (a) experiences that, despite being conscious, aren't re-represented during introspection; and (b) dissociated experiences inaccessible

---

1 At the time of this writing, the *Scientific American* essay was freely available online at: https://blogs.scientificamerican.com/observations/consciousness-goes-deeper-than-you-think/.

to the executive ego. If consciousness is inherent to all mentation, it may be fundamental in nature, as opposed to a product of particular types of brain function.

9.2 Introduction

The foundational theoretical inference of the clinical approach called 'depth psychology'—whose origins can be traced back to the works of Frederic Myers, Pierre Janet, William James, Sigmund Freud and Carl Jung—is that the human psyche comprises two main subdivisions: a conscious and an unconscious segment (Kelly et al. 2009: 301-334). The conscious segment comprises mental activity to which one has introspective access. The so-called 'ego' is the felt sense of personal self that arises in association with a subset of this introspectively-accessible activity—e.g. some bodily sensations, images, thoughts, beliefs, etc.—and it is in this sense that I use the word 'ego' throughout this paper. In contrast, the unconscious segment comprises mental activity to which one has no introspective access. Inaccessible as it may be, depth psychologists contend that mental activity in the 'unconscious'—a term often used as a noun—still can and does influence one's conscious thoughts, feelings and behaviors. A more modern articulation of the notion of a *mental unconscious*—as opposed to what has historically been called "unconscious cerebration" (Kelly et al. 2009: 340-352)—can be found in the writings of Kihlstrom (1997), for example.[2]

---

2 Throughout this volume, I use the word 'mental' as a synonym of 'phenomenal'; *except in this chapter*. Because this chapter was originally published as an article in a psychology journal, here the word 'mental' is associated with cognitive activity, instead of qualia. According to this definition, mental processes aren't necessarily conscious, for cognition can conceivably take place unconsciously. And if they are conscious, mental processes then entail the acquisition of knowledge and/or understanding, which implies more than just the presence of phenomenality.

Recent empirical results seem to corroborate the hypothesis of a mental unconscious by revealing the presence of mental activity individuals cannot access through introspection, but which nonetheless causally conditions the individuals' conscious thoughts, feelings and behaviors (e.g. Westen 1999, Augusto 2010, Eagleman 2011). Hassin goes as far as insisting, "unconscious processes can carry out every fundamental high-level function that conscious processes can perform" (2013: 196). He reviews empirical evidence indicating that the unconscious is capable of cognitive control, the pursuit of goals, information broadcasting and even reasoning (Hassin 2013: 197-200). This echoes Dijksterhuis and Nordgren, whose experiments indicate that the unconscious can encompass "all psychological phenomena associated with thought, such as choice, decision making, attitude formation and attitude change, impression formation, diagnosticity, problem solving, and creativity" (2006: 96). Even practitioners of cognitive therapy, who have traditionally ignored the unconscious, have more recently found clinical value in interpreting possible indirect manifestations of inaccessible mental activity in the form of dreams (Rosner, Lyddon & Freeman 2004). This new scientific approach to the hypothesis of an unconscious has been called "the new unconscious" (Hassin, Uleman & Bargh 2005).

Clearly, there is significant evidence for the presence of causally-effective mental activity that we ordinarily cannot access through introspection. The question, however, is whether mental activity inaccessible through introspection is necessarily *unconscious*. It is true that, from the perspective of clinical psychology, these two modalities are operationally indistinguishable, since the clinicians' sole gauge of their patients' range of consciousness is the patients' own introspective reports. However, from a theoretical standpoint, it is conceivable that mental activity the ego cannot access through introspection could still be conscious, in the sense of being phenomenally experienced somewhere in the psyche. If so, this has significant implications for our understanding of the nature of consciousness—and of its relationship to brain function—in the fields of neuropsychology, neuroscience and philosophy of mind.

Indeed, although the conflation between lack of introspective access and lack of consciousness is operationally justifiable in a clinical setting, the widespread use of the qualifier 'unconscious' today suggests an intrinsic dichotomy in the nature of mental processes: some supposedly *aren't* experienced whilst others, somehow, are. This implies that consciousness is not fundamental to mentation, but a property that emerges from particular arrangements or configurations of neurons. Primed and driven by this assumption, significant resources are spent in neuropsychology and neuroscience today in an effort to figure out what these arrangements or configurations are. Hypotheses currently under investigation vary from vast topologies of information integration across neurons (Tononi 2004) to microscopic quantum processes within neural microtubules (Hameroff 2006).

The present paper, on the other hand, elaborates on the possibility that these efforts are misguided, for introspectively-inaccessible mental processes may still be conscious: they may be phenomenally experienced in a manner—or in a segment of the psyche—that escapes egoic introspection. This way, the notion of an unconscious, despite the broad use and influence of the term in today's psychology, may at root be a linguistic inaccuracy originating from mere operational convenience. If so, then consciousness may not be the product of specific arrangements or configurations of neural activity, but a fundamental property of *all* mentation. The implications of this possibility for neuropsychology, neuroscience and philosophy of mind are hard to overestimate.

9.3 Defining and gauging consciousness

Before we can meaningfully discuss unconsciousness—the alleged lack of consciousness—we must, of course, have clarity regarding the meaning of the word 'consciousness.' What does it mean to say that a mental process is conscious? In this paper, I shall use a rigorous definition well-accepted in neuropsychology, neuroscience and philosophy of mind: mental activity is *conscious* if, and only if, there is something—*anything*—it is like to have such mental activity in and of itself (Nagel 1974, Chalmers 2003). (A less rigorous but more easily understandable formulation of this definition is this: mental activity is conscious if there is something it *feels* like to have such mental activity in and of itself. The verb 'to feel,' however, is too ambiguous to be used in a rigorous definition, so philosophers of mind have reached consensus around the formulation I originally proposed above.) This way, if mental activity is *un*conscious, then there is nothing it is like to have such activity in and of itself, even if it, in turn, causes or influences conscious activity. Notice that this definition of consciousness honors our intuitive understanding of the word: you only consider yourself conscious right now because there is something it is like to be you while you read this paper. Otherwise, you would necessarily be unconscious.

To remain consistent with our intuitive understanding of words, I shall also say that mental activity corresponds to *experience* if, and only if, it is conscious. You experience reading this paper because you are conscious of it right now. If you were not, what sense would there be in saying that you experience it?

According to these definitions, higher-order thought (as defined in Schooler 2002: 340) is unnecessary for there to be consciousness. The presence of the mere qualities of raw experience—which philosophers of mind call *qualia*—is already sufficient for a mental process to be considered conscious. In this context, the categorization proposed by Schooler is helpful: he distinguishes between "non-conscious (unexperienced), conscious (experienced), and meta-conscious (re-represented)" mental processes (2002: 339). Only the latter entails higher-order thought.

Now notice that *direct* insight into one's conscious inner life is limited to those experiences one's ego can access through introspection and then report to self or others. In the words of Klein, "It is *only* in virtue of knowledge by acquaintance that we know our mental states. … Accordingly, the use of introspective reports as a reliable and informative source of information about mental states has seen a resurgence over the past few decades" (2015: 361, original emphasis). For this reason, the study of the Neural Correlates of Consciousness (NCCs) still largely consists in correlating objective measurements of neural activity with introspective assessments (Koch 2004): patterns of neural activity accompanied by reported experience are considered NCCs. Indeed, as Newell and Shanks recently wrote, "Whereas issues about how to define and measure awareness were once highly prominent and controversial, it now seems to be generally accepted that awareness should be operationally defined as reportable knowledge" (2014: 15).

The problem is that, as I shall shortly elaborate upon, for the subject's ego to access and report an experience there must be: (a) an associative link between the ego and the experience; and (b) a meta-conscious re-representation of the experience. Therefore, while subjects can report non-dissociated meta-conscious processes, *they fundamentally cannot distinguish between truly unconscious processes and conscious processes that simply aren't meta-conscious*, for *both* types are equally unreportable to self and others. This is an alarming conclusion, for much of the work indicating the presence of an unconscious is based on (the lack of) introspective reports of experience. The next two sections expand on all this.

In what follows, I shall assume that introspective reports are as good as "reliable, relevant, immediate, and sensitive" (Newell

& Shanks 2014: 3). This is charitable towards the hypothesis of an unconscious, for—as Newell and Shanks argued (2014)— much of the evidence behind this hypothesis can be attributed to methodological artifacts: delayed introspective assessments leading to impaired recall, experimenters not providing sufficient opportunity for subjects to report the introspective insights they actually have, cross-task confusion, etc. My goal is to show that, *even if* the research underpinning the existence of an unconscious were free of methodological artifacts, there would *still* be compelling reasons to posit that mental processes unaccompanied by introspective reports of experience can be conscious nonetheless.

9.4 Non-self-reflective experiences

To gain introspective access to an experience it is not enough to merely have the experience; we must also consciously know *that* we have it. After all, what introspective insight could we gain about an experience of which we are not explicitly aware? Schooler elaborates:

> Critical to both the centrality of the conscious/non-conscious distinction, and its equation with reportability, is the assumption that people are explicitly aware of their conscious experiences. However, this assumption is challenged when subjective experience is dissociated from the explicit awareness of that experience. Such dissociations demonstrate the importance of distinguishing between consciousness and 'meta-consciousness.' (2002: 339.)

The conscious knowledge *of* the experience—which comes *in addition* to the experience itself—is what Schooler calls a "re-representation":

> Periodically attention is directed towards explicitly assessing

the contents of experience. The resulting meta-consciousness involves an explicit *re-representation* of consciousness in which one interprets, describes, or otherwise characterizes the state of one's mind. (2002: 339-340, emphasis added.)

Although re-representation is necessary for introspection, it is largely absent, for instance, in dreams (Windt & Metzinger 2007). This demonstrates compellingly that mental activity does *not* need to be re-represented in order to be experienced—after all, who can seriously doubt that dreams are experienced?—but only to be introspectively accessed. During ordinary dreams we simply experience, without consciously knowing *that* we experience.

More formally, suppose that one has an experience X. To gain introspective access to X one must have conscious knowledge N of X. But N—the "re-representation"—is a separate experience in its own right. One experiences the *knowing of X* as a quality closely related to, but distinct from, X itself. N is not encompassed, entailed or implied by X. Indeed, Schooler highlights the fact that re-representations can even *mis*represent the original experiences:

Once meta-consciousness is triggered, translation dissociations can occur if the re-representation process misrepresents the original experience. Such dissociations are particularly likely when one *verbally* reflects on non-verbal experiences or attempts to take stock of ambiguous or subtle perceptual experiences. (2002: 340, emphasis added.)

To make these abstract considerations more concrete, consider your breathing right now: the sensation of air flowing through your nostrils, the movements of your diaphragm, the inflation and deflation of your lungs, etc. Were you *not* experiencing these sensations a moment ago, before I directed your attention to them?[3] Or were you just unaware *that* you were experiencing them all along? By directing your attention to these sensations, did I make them *conscious* or did I simply cause you to experience the *extra* quality of knowing *that* the sensations were conscious? Clearly, even waking experiences can occur without re-representation.

Re-representations are the product of a self-reflective *configuration of* consciousness, whereby the latter turns in upon itself so to objectify its own contents (Kastrup 2014: 104-110). In humans, this usually occurs through the use of "semiotic mediation" (Valsiner 1998), which is our ability to re-represent our experiences by *naming* them explicitly or implicitly. Gillespie gives an example: "In order to obtain dinner one must first name ... one's hunger ... This naming, which is a moment of self-reflection, is the first step in beginning to construct, semiotically, a path of action that will lead to dinner" (2007: 678).

Naturally, nothing prevents experiences from occurring outside the field of self-reflection—that is, occurring without being explicitly or implicitly named. Nixon, for instance, calls these "unconscious experiences" (2010: 216), which in my view is an oxymoron but illustrates the subtlety of the point. He lists several examples: blindsight (Stoerig & Cowey 1997), prosopagnosia (Sacks 1985), sleepwalking, post-hypnotic suggestion, etc. Indeed, the emergence of so-called "no-report paradigms" in contemporary neuroscience attests to the abundant presence of waking experiences that are unreportable because they fall outside the field of self-reflection (Tsuchiya et al. 2015, Vandenbroucke et al. 2014).

Moreover, *the neural activity patterns of the NCCs themselves* suggest circumstantially—yet compellingly—that many NCCs correspond merely to a self-reflective configuration

---

3 Notice that *attention* is required to explicitly assess an experience at a metacognitive—that is, self-reflective—level.

of consciousness. To see this, notice first that the conscious knowledge *N* of an experience *X* is triggered by the occurrence of *X*. For instance, it is the occurrence of a sense perception that triggers the realization that one is perceiving something. *N*, in turn, evokes *X* by directing attention back to it: the realization that one is perceiving something naturally shifts one's mental focus back to the original perception. So we end up with a back-and-forth cycle of evocations whereby *X* triggers *N*, which in turn evokes *X*, which again triggers *N*, and so forth. See Figure 9.1 for an illustration.

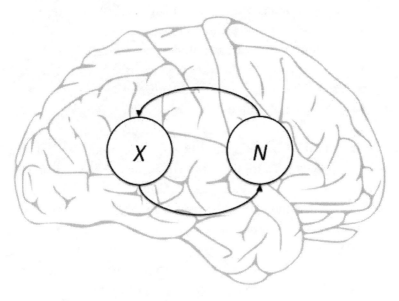

Figure 9.1: Illustrative caricature of oscillatory evocations between an experience (*X*) and the meta-conscious knowledge of the experience (*N*).

As it turns out, recent characterizations of the NCCs show precisely this pattern of reverberating back-and-forth communications between different brain regions (Dehaene & Changeux 2011, Boly et al. 2011, van Gaal et al. 2011). When damage to the primary visual cortex presumably interrupts this reverberation, patients display blindsight (Paller & Suzuki 2014: 387) — that is, the ability to correctly discriminate moving objects despite the *reported* inability to see them. This is precisely what one would expect if the reverberation in question were the oscillations between X and N: the objects *are* consciously perceived — therefore explaining how the patients can discriminate them — but the patients do not know *that* they consciously perceive the objects.

I thus submit that many NCCs are, in fact, the correlates only of a potentially very small subset of consciousness — namely, meta-consciousness or self-reflection — instead of consciousness proper. The introspectively inaccessible character of experience that isn't re-represented constitutes the first mechanism through which seemingly unconscious mental activity may, in fact, be conscious. There is yet another mechanism, which will be explored in the next section.

9.5 Dissociated experiences

Dissociative states are well recognized in psychiatry today, featuring prominently in the DSM-5 (American Psychiatric Association 2013). Their hallmark is "a disruption of and/ or discontinuity in the normal integration of consciousness, memory, identity, emotion, perception, body representation, motor control, and behavior" (Black & Grant 2014: 191). In other words, dissociation entails fragmentation of the contents of consciousness.

There are different forms of dissociation. Klein (2015), for instance, discusses a form in which the subject's ego loses the sense of ownership of some of the subject's own mental states. This occurs when consciousness can no longer "relate to its object in a particular, self-referential way" (Klein 2015: 362). He lists several examples, such as the case of a man who, after an accident, could accurately report the content of his memories but "was unable to experience that content as his own" (Klein

2015: 368). Notice, however, that the man's ego could *still* access the content; just not identify with it.

In what follows, I shall focus on a strong form of dissociation in which the ego *cannot even access* certain contents of consciousness. In its pathological variations, this is known as Dissociative Identity Disorder (DID). A person suffering from DID exhibits multiple, disjoint centers of consciousness called alters. Each alter experiences the world as a distinct personality (Braude 1995).

Although there has been debate about the authenticity of DID as a psychiatric condition—after all, it is conceivable that patients could fake it—research has confirmed DID's legitimacy (Kelly et al. 2009: 167-174 & 348-352). Two recent studies are particularly interesting to highlight. In 2015, doctors reported on the case of a German woman who exhibited a variety of alters (Strasburger & Waldvogel). Peculiarly, some of her alters claimed to be blind while others could see normally. Through EEGs, the doctors were able to ascertain that the brain activity normally associated with sight wasn't present while a blind alter was in control of the woman's body, even though her eyes were open. When a sighted alter assumed executive control, the usual brain activity returned. This is a sobering result that shows the literally *blinding* power of dissociation. In another study (Schlumpf et al. 2014), investigators performed functional magnetic resonance imaging (fMRI) brain scans on both DID patients and actors simulating DID. The scans of the actual patients displayed clear and significant differences when compared to those of the actors. Undoubtedly, thus, DID is real.

Normally, only one of the alters has executive control of the body at any given moment. The important question for the purposes of the present paper is then this: Can the *other* alters, who are *not* in control of the body, remain conscious or do they simply fade into unconsciousness? If they can remain conscious, the implication is that a person can have multiple *concurrent* but dissociated centers of *consciousness*, as originally hypothesized by Frederic Myers and Pierre Janet (Kelly et al. 2009: 305-317). Presumably, then, each center has its own private, parallel stream of experiences.

Occasionally, however, the dissociation isn't bilateral: a first alter is able to gain partial access to the experiences of a second, without the second alter being able to access the experiences of the first. This rare kind of unilateral dissociation provides tantalizing indications that alters can remain conscious even when not in control of the body. In Morton Prince's well-known study of the 'Miss Beauchamp case' of DID, one of the alters—called Sally—"was a co-conscious personality in a deeper sense. When she was not interacting with the world, she did not become dormant, but persisted and was active" (Kelly et al. 2009: 318). Sally maintained that she knew

> everything Miss Beauchamp ... does at the time she does it, — knows what she thinks, hears what she says, reads what she writes, and sees what she does; that she knows all this as a separate co-self, and that her knowledge does not come to her afterwards ... in the form of a memory. (Prince, as quoted in Kelly et al. 2009: 318)

Stephen Braude's more recent work reinforces the view that alters can be co-conscious "discrete centers of self-awareness" (1995: 67). He points—as evidence for this hypothesis—at the struggle of different alters for executive control of the body and the fact that alters "might intervene in the lives of others [i.e. other alters], intentionally interfering with their interests and activities, or at least playing mischief on them" (Braude 1995: 68). It thus appears that alters can not only be concurrently conscious, but that they can also vie for dominance with each other.

Strong dissociation is not restricted to DID—its extreme

form—or to pathology, for that matter. Indeed, the foundational hypothesis of depth psychology entails a form of natural dissociation between the conscious ego and the so-called "unconscious." As such, it is plausible—in fact, there is overwhelming clinical evidence for it in the annals of depth psychology—that we all have at least one dissociated mental subsystem that we cannot access through introspection. Ernest Hilgard (1977) conceived of these dissociated subsystems as conscious, much as Myers, Janet and Braude did.

Thus, the possibility that presents itself to us is that we may all have one or more *conscious* 'others' within ourselves, dissociated from our ego. If this is so, then (a) our ego ordinarily has no introspective access to the experiences of these 'others'; and, consequently, (b) the study of the NCCs is largely blind to the potentially idiosyncratic patterns of neural activity corresponding to such dissociated experiences. This is the second mechanism through which apparently unconscious mental activity may, after all, be conscious.

9.6 A model of dissociation

Wegner (2002) proposes an analogy for explaining alters: different operating systems running on the same hardware. This way, the transfer of executive control from one alter to another would be analogous to shutting down Windows and rebooting the computer with Linux. This, of course, only accounts for strictly alternating personalities and thus fails to explain much of the clinical data cited above. Nonetheless, it still suggests a starting point for a plausible model of dissociation.

If we define an *experiential frame* as the set of all qualities we experience at a given moment—encompassing our conscious perceptions, thoughts, emotions, bodily sensations, imagination, etc.—conscious life can be modeled as a chain of experiential frames. This is graphically illustrated in Figure 9.2, wherein experiential frames F1 to Fn are shown. Each frame is evoked by the previous frame through cognitive associations, in the sense that e.g. our particular thoughts in the present moment largely determine which emotions we experience in the next moment; or that our emotions in the present moment largely determine our actions—and therefore perceptions—in the next moment; and so on. These cognitive associations are represented by the arrows linking frames together in Figure 9.2.

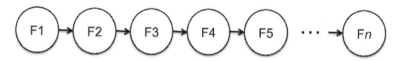

Figure 9.2: Conscious life as a chain of experiential frames connected through cognitive associations.

Wegner's suggestion can then be visualized as in Figure 9.3. The chain of experiential frames—denoted F—corresponding to a first alter is interrupted by experiential frames—denoted F'—corresponding to a second alter. The key point is that, once executive control is assumed by the experiential frames F' of the second alter, the corresponding experiential frames F of the first alter *cease to exist*. There is no parallelism of experience: either the mental contents of the first alter are experienced or those of the second alter; never those of both concurrently. As such, this is a *sequential model of dissociation* and, as we've seen, it isn't sufficient to explain the clinical data cited.

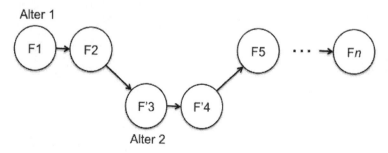

Figure 9.3: The sequential model of dissociation in the context of DID.

Alternatively, we can hypothesize that the chains of experiential frames of *both* alters are *always* present, concurrently and in parallel. Executive control of the body simply switches between the two parallel chains, as shown in Figure 9.4. Experiential frames drawn in grey represent those without executive control, *but still conscious*. This is thus a *parallel model of dissociation*, which illustrates the hypothesis of "co-consciousness" (a term originally coined by Morton Prince, as discussed by Kelly et al. 2009: 317).

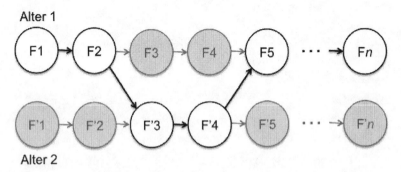

Figure 9.4: The parallel model of dissociation in the context of DID.

We have seen that DID is a pathological form of dissociation, but that we may all naturally have strongly dissociated mental subsystems that never—or very seldom—vie for executive control of the body. These would constitute the so-called "unconscious" of depth psychology. Figure 9.5 illustrates how such strongly dissociated mental subsystems can be modeled under the proposed framework. For simplicity, only the ego and one dissociated subsystem are shown. The 'other' in this case—represented by the dissociated chain of experiential frames F'—is content to live its inner life in the background of egoic activity. It only manifests its presence through indirect, subtle influences on egoic experiences, as represented by the dashed arrows vertically linking the two chains. These subtle influences can take many forms, such as: dissociated emotions influencing our egoic thoughts and behaviors (Lynch & Kilmartin 2013: 100); dissociated beliefs and expectations influencing our egoic perceptions (Eagleman 2011: 20-54); dissociated drives manifesting themselves symbolically in the form of dreams (von Franz & Boa 1994, Jung 2002, Fonagy et al. 2012); etc.

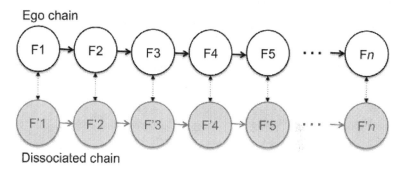

Figure 9.5: The parallel model of dissociation in a depth-psychological context.

Admittedly, limitations in our ability to gauge consciousness currently prevent us from asserting with certainty, on an empirical basis, that the parallel model of dissociation is correct. However, by the same token, we can also not assert that it isn't. The brain seems to have sufficient resources for this kind of parallelism and, if anything, the clinical data is suggestive of its validity (again, Kelly et al. 2009: 305-322 and Braude 1995). The parallel model should, therefore, be considered not only plausible but perhaps even probable, in which case it further substantiates the notion that the "unconscious" may be—well—conscious.

9.7 Discussion

I have elaborated on the hypothesis that there may be no such a thing as an unconscious mental process. All mental processes may be conscious, in the sense that there may be something it is like to have such mental processes in and of themselves. Our impression that some mental processes are unconscious may arise from (a) their consisting in non-self-reflective experiences not amenable to introspection or (b) their being strongly dissociated from the executive ego and, therefore, inaccessible to it.

Underlying this entire paper is the differentiation between consciousness proper and particular *configurations of* consciousness, such as self-reflection and dissociative states. It is rather disturbing how often these notions are conflated not only in general psychology, but also in neuroscience and philosophy of mind. For instance, a relatively recent article (Gabrielsen 2013) talks about the emergence of consciousness in human babies when what is discussed is—as per the argument developed in this paper—likely to be the emergence of meta-consciousness.[4]

Dijksterhuis and Nordgren also "define conscious thought as object-relevant or task-relevant cognitive or affective thought processes that occur while the object or task is *the focus of one's conscious attention*" (2006: 96, emphasis added). They insist, "it is very important to realize that *attention is the key* to distinguish between unconscious thought and conscious thought. *Conscious thought is thought with attention*" (Dijksterhuis & Nordgren 2006, emphasis added). In appealing to *attention*, as opposed to experience or *qualia*, they are implicitly associating consciousness with self-reflection or re-representation, as discussed in Section 9.4.

Even more strikingly, Cleeremans (2011) *explicitly defines* consciousness as self-reflection. He overtly conflates experience with meta-consciousness and reportability:

Awareness, on the other hand, always seems to minimally entail the ability of knowing *that* one knows. This ability, after all, forms the basis for the verbal reports we take to be the most direct indication of awareness. And when we observe the

---

4 For clarity, by "emergence of meta-consciousness" I mean here the early, or even precursor, stages of meta-consciousness.

absence of such ability to report on the knowledge involved in our decisions, we rightfully conclude that the decision was based on unconscious knowledge. Thus, it is when an agent exhibits *knowledge* of the fact that he is sensitive to some state of affairs that we take this agent to be a conscious agent. This *second-order* knowledge, I argue, critically depends on *learned* systems of meta representations, and forms the basis for conscious experience. (Cleeremans 2011: 3)

This isn't a recent problem. When one reads the original texts of the founders of depth psychology whilst holding the distinction between consciousness and meta-consciousness in mind, one quickly realizes that, when they spoke of unconsciousness, the founders often meant a lack of *meta*-consciousness—not of experience proper. This is abundantly evident, for instance, in an essay written by Carl Jung in the 1920s or early 1930s, called "The Stages of Life" (Jung 2001: 97-116).

It could be argued that the distinction between experience and meta-consciousness is merely a semantic point. However, consider this: by conflating consciousness proper with *self-reflective* consciousness, we also indirectly equate non-self-reflective consciousness with unconsciousness; we absurdly imply that dreams—which largely lack self-reflection (Windt & Metzinger 2007)—aren't experienced. Instead of the three categories proposed by Schooler—namely, "non-conscious (unexperienced), conscious (experienced), and meta-conscious (re-represented)" (2002: 339)—we are left with only two: non-conscious and meta-conscious. Consequently, we are forced to collapse the conscious onto the non-conscious and, in the process, end up disregarding the extraordinary phenomenon of *qualities of experience*.[5] Clearly, this isn't merely semantic.

---

5 That is, we end up sweeping the "hard problem of consciousness" (Chalmers 2003) under the rug.

Most importantly, the philosophical implications of mistaking consciousness for *meta*-consciousness are significant. If some mental processes were truly unconscious while others are conscious, it would follow that consciousness is the product of some specific anatomical and/or functional arrangements of brain activity. In other words, consciousness would be derivative, as opposed to fundamental. Philosophically, this would corroborate the ontology of physicalism (Stoljar 2016) while contradicting alternatives like panpsychism (Strawson et al. 2006), cosmopsychism (Shani 2015) and idealism (Kastrup 2017b[6]). It would leave us with no way to circumvent the arguably insoluble "hard problem of consciousness" (Chalmers 2003).

On the other hand, if consciousness is inherent to all mental processes, then the specific anatomical and/or functional parameters of different processes correspond merely to different *contents and/or configurations of consciousness*—that is, to the particular qualities that are experienced—but do not determine the presence or absence of consciousness itself. This allows us to circumvent the "hard problem of consciousness" altogether, by inferring that consciousness is primary. While it's not my intent in this paper to argue for or against any particular ontology of mind, it is significant that a lucid, critical interpretation of the available empirical data leaves more avenues of philosophical inquiry open.

If we are true to the spirit of the words 'consciousness' and 'experience,' diligent in our interpretation of empirical observations—both experimental and clinical—and rigorous in our use of concepts, we are led not only to the conclusion that *all* mental processes may be conscious, but that consciousness itself may be fundamental.

---

6 This reference can be found in Chapter 5 of the present volume.

Part IV

Neuroscientific evidence

That which is usually held to be a greater complexity of the psychical state appears to us ... to be a greater dilatation of the whole personality, which ... expands with the unscrewing of the vice in which it has allowed itself to be squeezed, and, always whole and undivided, spreads itself over a wider and wider surface. That which is commonly held to be a disturbance of the psychic life itself, an inward disorder, a disease of the personality, appears to us ... to be an unloosing or a breaking of the tie which binds this psychic life.
Henri Bergson: *Matter and Memory.*

To formulate and express the contents of this reduced awareness, man has invented and endlessly elaborated those symbol-systems and implicit philosophies which we call languages. Every individual is at once the beneficiary and the victim of the linguistic tradition into which he or she has been born ... the victim in so far as it confirms him in the belief that reduced awareness is the only awareness and as it bedevils his sense of reality, so that he is all too apt to take his concepts for data, his words for actual things.
Aldous Huxley: *The Doors of Perception.*

Chapter 10

Preamble to Part IV

As discussed earlier, the main line of empirical evidence against mainstream physicalism today probably comes from laboratory experiments corroborating the quantum mechanical prediction of contextuality. These experimental results have been discussed in Chapter 6, are elaborated upon more carefully and extensively in Chapter 15, and finally recapitulated in the Appendix.

Nonetheless, this Part IV focuses solely on *neuroscientific*— not physical—empirical evidence. The reason for this choice is two-fold: firstly, the microscopic realms of quantum mechanics are considered by many—even scientists and philosophers—too remote to base a compelling empirical case against physicalism. After all, when we look around we don't see a quantum mechanical world, but a classical one, wherein objects seem to have definite position and momentum. Since we know that reality ultimately is quantum mechanical in nature, strictly speaking this reasoning is just wrong. Yet, I have chosen to acquiesce to it here and focus on lines of evidence whose relevance cannot be dismissed. Secondly, more than just *prima facie* contradicting physicalism, the empirical evidence from neuroscience also *directly suggests idealism*. In the context of a book that, more than just rejecting physicalism, argues *for* idealism, this seems to make for a more compelling case.

Chapter 11 compiles and discusses a surprisingly broad list of instances of brain function *impairment* that are accompanied by *enrichment* of conscious inner life and an *expansion* of one's sense of identity. The list includes cases as varied as asphyxiation, cardiac arrest, physical trauma to the head, the consumption of psychoactive substances that dampen brain activity, and many others. In all these cases, subjects report experiences and insights that surpass the ordinary envelop of possibilities.

Such correlations between *impaired* brain function and *enriched* conscious inner life are at least cumbersome to explain under the physicalist postulate that conscious inner life is constituted or generated by brain activity (Chapter 11 substantiates this claim more thoroughly). Under idealism, on the other hand, they are entirely expected: if normal brain function is the extrinsic appearance of *dissociated* consciousness, then a *reduction* or *impairment* of normal brain function should be the extrinsic appearance of a *reduction* or *impairment of the dissociation*. And, of course, from a first-person perspective a reduction of dissociation can only be experienced as an enrichment of inner life: reintegrated memories, the recovery of a broader sense identity, renewed access to previously dissociated insights and emotions, reintegration of previously dissociated skills, etc. Contrary to physicalism, idealism can thus not only accommodate, but also make sense of, the evidence discussed in Chapter 11.

Naturally, the argument in Chapter 11 is not that *all* impairment of brain function should be accompanied by enriched inner life. Otherwise, the smartest, most creative and most spiritually enlightened people would be those with the most damaged brains. This is clearly not the case. *But neither does idealism require it to be the case.* Allow me to elaborate.

As we have seen in Part II, a living organism corresponds to a dissociated alter of universal consciousness. As such, each person can be regarded as a segment of universal consciousness—meant here generically, without implying that universal consciousness necessarily has spatial or temporal extension—comprising its own dissociated mental contents. Segments comprising lots of mental contents can be referred to as 'big alters,' whereas segments comprising few mental contents can be referred to as 'small alters.' It is reasonable to speculate, for instance, that human beings correspond to bigger alters than, say, insects.

Notice that *both big and small alters can be equally well*

dissociated. In other words, the relative amount of mental contents encompassed by an alter does not bear relevance to how well dissociated these mental contents are from the rest of universal consciousness. It is entirely coherent, within the logic of idealism, that a small alter could be more strongly dissociated from universal consciousness than a big alter, or the other way around.

Now, since brain activity is part of the extrinsic appearance of an alter's dissociated mental contents, it stands to reason that some—even *most*—types of brain function impairment will correspond simply to a diminution of the mental contents of the alter. These types of brain function impairment will *not* disrupt the dissociation itself, but only stifle whatever is circumscribed by the dissociative boundary. The alter will become smaller, cognitively compromised, but still equally well dissociated. This is why, under idealism, many or even most types of brain function impairment should still come accompanied by cognitive deficit, not awareness expansion.

Only some specific types of brain function impairment, which somehow affect the dissociative mechanisms themselves—as opposed to the mental contents encompassed by the alter— should correlate with an enrichment of conscious inner life. They make the dissociative boundary 'porous,' so to speak. At present, however, it is not yet known what precise aspects of brain function correspond to these dissociative mechanisms, even though some tantalizing indications are discussed in the next two chapters. For this reason, it is still impossible to predict with accuracy what types of brain function impairment should lead to what type of effect—awareness expansion or cognitive deficit.

What distinguishes the predictions of idealism from those of physicalism is this: under idealism, at least *some* types of brain function impairment *should* lead to enriched conscious inner life. Under physicalism, however, this is much more difficult to argue, as I elaborate upon in the next two chapters.[1] Strictly worded, my claim is therefore this: *there are certain types of brain function impairment—which under physicalism should correlate with cognitive deficit and under idealism with enriched inner life—that have been shown to be accompanied by enriched inner life. This corroborates idealism and contradicts physicalism.* In the next two chapters, I will attempt to substantiate this claim with neuroscientific empirical evidence.

Notice that, because of the inherent limitations we face in gauging consciousness—as discussed in Chapter 9—the playing field here isn't level: many types of brain function impairment may cause both an enrichment of conscious inner life *and* compromise the subjects' ability to *report* this enrichment. For instance, language or motor centers, memory pathways, or a variety of other communication-critical functions in the brain may be compromised, harming or eliminating the subjects' ability to speak or write. Many subjects could be lying in hospital with severe head trauma or other brain ailments, having unfathomable inner experiences, and yet be utterly incapable to relate any of it to family or medical staff. If brain areas essential to metacognition are compromised, subjects may not even be able to report their experiences *to themselves*, as also discussed in Chapter 9. Consequently, the evidence that corroborates idealism is restricted by contradicting requirements: the corresponding brain function impairment must sufficiently affect dissociative mechanisms—not just dampen the mental contents encompassed by the alter—whilst sufficiently preserving cognitive function so subjects can report their expanded awareness. These contradictory requirements aren't at all trivial to meet concurrently and, in fact, it is rather surprising that so many case reports exist in the literature that

---

1 I am well aware of the role of inhibitory mechanisms in the brain and discuss them at length in the next two chapters.

corroborate idealism.

One more point must be made. Under ideal circumstances, subjects should be fully instrumented and lying inside a functional MRI scanner (or any other apparatus capable to dynamically measure brain activity with sufficient temporal and spatial resolution) before, during and after the occurrence of brain function impairment. At the same time, subjects should be continuously reporting their conscious inner state according to standardized procedures. This would allow researchers to precisely correlate subjective reports with changes in the subjects' brain function profile, so firm conclusions could be extracted regarding the relationship between the two.

In reality, however, this ideal set of circumstances is almost never possible to achieve. No researcher, for instance, can deliberately induce cardiac arrest in a subject after having placed him or her inside an fMRI scanner. For this reason, the vast majority of reports of enriched conscious inner life following cardiac arrest are not accompanied by detailed scans of the subjects' brain activity profile. Similarly, although the literature contains many reports of awareness expansion during G-Force-induced Loss of Consciousness (G-LOC), it is impossible to properly instrument the subject while he or she is sitting inside a centrifuge.

There is one exception, though: research on the effects of psychedelic substances has been carried out under the ideal set of circumstances described above. Fully instrumented subjects were placed inside functional MRI scanners, instructed to report on their conscious inner state according to standardized procedures, and then injected with the psychedelic compound. Their brain activity profile was measured before, during and after the effects of the psychedelic and then correlated with their subjective reports. As it turns out, psychedelics have been shown to reduce brain activity levels (details ahead).

For this reason, brain-imaging studies of the effects of

psychedelics are the most fruitful and reliable as far as our ability to extract conclusions from cases of brain function impairment is concerned. I have thus dedicated Chapter 12 to a philosophical analysis of their results.

More than just confirming the pattern already discussed in Chapter 11, Chapter 12 offers a thorough analysis of the implications of physicalism regarding the relationship between brain activity and subjective experience. While acknowledging that—because it is so vaguely defined in this regard—physicalism can theoretically accommodate a variety of different mappings between brain activity profiles and reported inner life, it has one inescapable implication: whatever the specific mapping, *under physicalism all information discernible in a subject's conscious inner life must correspond to information measurable—at least in principle—in the subject's brain metabolism.* Otherwise, at least some of the information experienced by the subject would have no grounding in brain metabolism. This kind of disembodied information is incompatible with the spirit of physicalism, whatever particular mapping between brain activity profile and experience is postulated.

Explicating this key implication of physicalism is a central philosophical contribution of Chapter 12. Nonetheless, since we are here concerned with empirical evidence, it goes further: the ideal circumstances under which psychedelic brain-imaging studies were carried out enabled me to assess whether their results violated this key implication of physicalism. And indeed, they seem to have done so. This shifts the explanatory burden to the physicalist camp: it is up to them now to somehow reconcile the psychedelic studies' results with their ontology.

As things stand, I believe the next two chapters provide compelling evidence that idealism fits better with the available neuroscientific data than physicalism.

Chapter 11

Self-transcendence correlates with brain function impairment

This article first appeared in the *Journal of Cognition and Neuroethics*, ISSN: 2166-5087, Vol. 4, No. 3, pp. 33-42, in January 2017. The *Journal of Cognition and Neuroethics* is published by the Center of Cognition and Neuroethics, a joint venture between the Philosophy Department at the University of Michigan-Flint and the Insight Institute of Neurosurgery and Neuroscience. A summary of this article has also appeared in *Scientific American* on 29 March 2017.[1]

11.1 Abstract

A broad pattern of correlations between mechanisms of brain function impairment and self-transcendence is shown. The pattern includes such mechanisms as cerebral hypoxia, physiological stress, transcranial magnetic stimulation, trance-induced physiological effects, the action of psychoactive substances and even physical trauma to the brain. In all these cases, subjects report self-transcending experiences often described as 'mystical' and 'awareness-expanding,' as well as self-transcending skills often described as 'savant.' The idea that these correlations could be rather trivially accounted for on the basis of disruptions to inhibitory neural processes is reviewed and shown to be implausible. Instead, this paper suggests that an as-of-yet unrecognized causal principle underlying the entire pattern might be at work, whose further elucidation through

---

1 At the time of this writing, the *Scientific American* essay was freely available online at: https://blogs.scientificamerican.com/guest-blog/transcending-the-brain/.

systematic research could hold great promise.

11.2 Introduction

In this paper, 'self-transcendence' is defined as the abrupt—thus *not* gradual— broadening of one's sense of self through a step-function enrichment of one's subjective inner life. This can happen, for instance, when one suddenly acquires (a) a feeling that one is no longer confined to the spatio-temporal locus of the physical body; (b) entirely new mental skills that one has never attempted to develop through learning or training; or (c) unfamiliar emotions, insights or inner imagery. This essay attempts to show that there is a consistent pattern of correlations between self-transcendence—so defined—and a broad variety of brain function impairment mechanisms. In other words, several types of brain function *impairment* are consistently accompanied by *richer* inner life. This is counterintuitive and suggests a common underlying causal principle yet to be understood in its full scope.

In the next sections, several mechanisms of brain function impairment and the resulting self-transcendence effects will be reviewed. The goal is to establish a broad pattern by highlighting the similarities of the mechanisms and their effects.

11.3 Cerebral hypoxia

Fainting or near-fainting caused by restrictions of oxygen supply to the brain is known to induce liberating feelings of self-transcendence. For instance, the potentially fatal 'choking game' played by teenagers worldwide (Macnab 2009) is an attempt to induce such feelings through partial strangulation (Neal 2008: 310-315). The psychotherapeutic technique of holotropic breathwork (Rhinewine & Williams 2007), as well as more traditional yogic breathing practices, use hyperventilation to achieve similar effects: by increasing blood alkalinity levels, they interfere with normal oxygen uptake in the brain and ultimately lead to what is described as an expansion of ordinary awareness (Taylor 1994). Even straightforward hyperventilation outside a therapeutic context can lead to self-transcending experiences, such as described in this anecdotal—though representative— report:

> One of us stood against a tree and breathed deeply for a while and then took a very deep breath. Another pushed down hard on his ribcage ... This rendered the subject immediately unconscious ... When I tried it, I didn't think it would work, but then suddenly I was in a meadow which glowed in yellow and red, everything was extremely beautiful and funny. This seemed to last for ages. I must say that I have never felt such bliss ever again. (Retz 2007)

Finally, pilots undergoing G-force induced Loss of Consciousness (G-LOC)—whereby blood is forced out of the brain, causing hypoxia—report "memorable dreams" phenomenologically similar to near-death experiences (Whinnery & Whinnery 1990), which are notoriously self-transcending in character.

11.4 Generalized physiological stress

Near-Death Experiences (NDEs) are the prime examples of self-transcendence associated with dramatically reduced brain function due to e.g. cardiac arrest (van Lommel 2001). They reportedly entail life-transforming phenomenality— encompassing insights, emotions and rich inner imagery—far surpassing the envelop of ordinary experiences (Kelly et al. 2009: 367-421), despite overwhelming disruption to the brain's ability to operate. A recent and well-publicized NDE, which occurred while the patient was under close supervision of medical staff, captures this self-transcendent dimension. In the patient's own words:

I certainly don't feel reduced or smaller in any way. On the contrary, I haven't ever been this huge, this powerful, or this all-encompassing. ... [I] felt greater and more intense and expansive than my physical being. (Moorjani 2012: 69)

In a related manner, traditional initiatory rituals in pre-literate cultures sought to reveal the true nature of self and world through physical ordeals (Eliade 2009). It is reasonable to imagine that these ordeals—such as long sessions in sweat lodges, exposure to the elements, extreme exertion and even poisoning—physically compromised brain function through generalized physiological stress, thereby inducing self-transcending experiences.

11.5 Electromagnetic impairment

The use of transcranial magnetic stimulation can inhibit activity in localized areas of the brain by impairing the associated electromagnetic fields. As reported in a study (Blanke 2002), when neural activity in the angular gyrus of a patient with epilepsy was inhibited in this way, self-transcending out-of-body experiences were induced.

11.6 Trance-induced impairment

During the practice of so-called 'psychography,' an alleged medium enters a trance state and writes down information allegedly originating from a transcendent source beyond the medium's ordinary self. A neuroimaging study (Peres 2012) revealed that experienced mediums displayed marked reduction of activity in key brain regions—such as the frontal lobes and hippocampus—when compared to regular, non-trance writing. Despite this, text written under trance scored consistently higher in a measure of complexity than material produced without trance. As an observant science journalist remarked, more complex writing typically would require more activity in the frontal and temporal lobes—but that's precisely the opposite of what was observed. To put this another way, the low level of activity in the experienced mediums' frontal lobes should have resulted in vague, unfocused, obtuse garble. Instead, it resulted in more complex writing samples than they were able to produce while not entranced. Why? No one's sure. (DiSalvo 2012)

11.7 Chemical impairment

Psychedelic substances have been known to induce powerful self-transcending experiences (Strassman 2001, Griffiths et al. 2006, Strassman et al. 2008). It had been assumed that they did so by exciting parts of the brain. Yet, recent neuroimaging studies have shown that psychedelics do largely the opposite (Carhart-Harris et al. 2012, Palhano-Fontes et al. 2015, Carhart-Harris et al. 2016).[2] In an article he wrote for *Scientific American Mind*, neuroscientist Christof Koch (2012) expressed his surprise at these results. Carhart-Harris (2012: 2138), for instance, reported "only decreases in cerebral blood flow" under the influence of a psychedelic. Perhaps even more significantly, "the magnitude of this decrease [in brain activity] predicted the intensity of the subjective effects" of the psychedelic (Ibid.). As such, the significant self-transcending experiences that follow psychedelic intake are—counterintuitively—accompanied by reductions of brain activity.

11.8 Physical damage

If the trend above is consistent, we should expect some types of

---

2 A later study performed at the University of Zürich has confirmed this further, showing that a psychedelic causes "significantly reduced absolute perfusion" (that is, blood flow) in just about every region of the brain, whilst leading to "profound subjective drug effects" (Lewis et al. 2017).

physical brain damage to also correlate with self-transcending experiences. And indeed, this has been reported. In a recent study (Cristofori 2016), CT scans of more than one hundred Vietnam War veterans showed that damage to the frontal and parietal lobes increased the likelihood of self-transcending "mystical experiences." In a previous study (Urgesi et al. 2010), patients were evaluated before and after brain surgery for the removal of tumors, which caused collateral damage in surrounding tissue. Statistically significant increases in feelings of self-transcendence were reported after the surgery.

The self-transcending character of experiences that accompany certain types of brain injury has been evocatively described by neuroanatomist Jill Bolte Taylor, following a stroke that damaged her brain's left hemisphere:

> my perception of my physical boundaries was no longer limited to where my skin met air. I felt like a genie liberated from its bottle. The energy of my spirit seemed to flow like a great whale gliding through a sea of silent euphoria. (Taylor 2009: 67)

The similarity to Moorjani's experience quoted earlier (2012: 69) is striking, despite the latter having been caused by generalized physiological stress, not a left-hemisphere stroke.

Not only 'mystical experiences' correlate with brain damage, but also the emergence of new mental skills. The literature reports many cases of so-called 'acquired savant syndrome,' wherein an accident or disease leading to brain injury gives rise to genius-level abilities (Lythgoe et al. 2005, Treffert 2006, Treffert 2009: 1354, Piore 2013). There are examples of such abilities arising after meningitis, bullet wounds to the head, and even with the progression of dementia (Miller et al. 1998, 2000).

The Idea of the World

11.9 Discussion

As we've seen, there is a broad pattern associating a variety of brain impairment mechanisms with self-transcending experiences. A potential explanation for this is that brain function impairment could disproportionally affect inhibitory neural processes, thereby generating or bringing into awareness other neural processes associated with self-transcending experiences. There are, however, problems with this explanation.

Under the physicalist assumption that experience is constituted or generated by brain activity, an increase in the richness of experience—as often entailed by self-transcendence—must be accompanied by an increase in the metabolism associated with the neural correlates of experience (Kastrup 2016b[3]). This is so because (a) there supposedly is nothing to experience but its neural correlates; and (b) richer experience spans a broader information space in awareness that only increased metabolism can create in the physical substrate of the brain. Any other alternative would decouple experience from the workings of the living brain information-wise, contradicting physicalism. As such, it is difficult to see why a reduction of oxygen supply to the brain *as a whole*—as in partial strangulation, hyperventilation, G-LOC, cardiac arrest, etc.—would selectively affect inhibitory neural processes, while maintaining enough oxygen supply to feed an increase in the neural correlates of experience.

Alternatively, one could speculate that self-transcending experiences occur only *after* normal brain function resumes, subsequent to e.g. restoration of oxygen supply. This, however, cannot account for several of the cases reviewed above. For instance, during the neuroimaging studies of the psychedelic state (e.g. Carhart-Harris et al. 2012) researchers collected subjective reports of self-transcendence while *concurrently* monitoring the subjects' reduced brain activity levels. The same holds for the

---

3 See Chapter 12, where the cited paper is reproduced.

neuroimaging study of psychography (Peres 2012). Similarly, in the case of acquired savant (e.g. Treffert 2006, Treffert 2009: 1354) new mental skills are also *concomitant* with the presence of physical damage in the brain. And even in the case of NDEs, there are arguments for why confabulation after resumption of normal brain function cannot account for some of the reported experiences (Kelly et al. 2009: 419-421).

Appeals to impairment of inhibitory processes to explain acquired savant syndrome are particularly difficult to defend. They necessarily entail that the savant skills are pre-developed in the brain but remain inhibited. Brain function impairment occasioned by e.g. trauma then supposedly unlocks these dormant skills by shutting down inhibitory processes. One must wonder, however, how the brain could have developed extraordinary skills, such as e.g. prodigious aptitude for calculations, without any training. And if these skills—many of which are advantageous for survival—were latent in us all, why would the brain have evolved to keep them inhibited in the first place?

It is conceivable that individual cases of self-transcendence could have their own idiosyncratic explanation, unrelated to the other cases, and that the overall pattern suggested in this paper is a red herring. For instance, one could tentatively explain (a) the euphoric effects of hypoxia by speculating that it e.g. somehow triggers the brain's reward system, while accounting for (b) the expansion of one's sense of identity beyond the physical body— as reported by Taylor (2009: 67)—through e.g. damage to the orientation association area of the left brain hemisphere. But given the sometimes-striking similarities in the phenomenality reported across the cases reviewed and the fact that *all* cases— despite their different mechanisms of action—entail impairment of brain function, the question is whether it is plausible that no common causal principle is at work.

The current data is at least suggestive of a single, yet-unrecognized causal principle underlying all cases. More systematic studies of the subjective effects of brain function impairment—leveraging e.g. psychedelic compounds and trans-cranial magnetic stimulation—in specific brain regions could help unveil this principle. Could one e.g. reliably trigger savant skills or mystical experiences by inhibiting neural activity in particular areas under controlled conditions? What would the implications of such a scenario be? Questions such as these hold not only great public interest, but also high significance for both neuroscience and neurophilosophy.

Chapter 12

What neuroimaging of the psychedelic state tells us about the mind-body problem

This article first appeared in the *Journal of Cognition and Neuroethics*, ISSN: 2166-5087, Vol. 4, No. 2, pp. 1-9, in July 2016. The *Journal of Cognition and Neuroethics* is published by the Center of Cognition and Neuroethics, a joint venture between the Philosophy Department at the University of Michigan-Flint and the Insight Institute of Neurosurgery and Neuroscience.

12.1 Abstract

Recent neuroimaging studies of the psychedelic state, which have commanded great media attention, are reviewed. They show that psychedelic trances are consistently accompanied by broad reductions in brain activity, despite their experiential richness. This result is at least counterintuitive from the perspective of mainstream physicalism, according to which subjective experience is entirely constituted by brain activity.[1] In this brief analysis, the generic implications of physicalism regarding the relationship between the richness of experience and brain activity levels are rigorously examined from an informational perspective, and then made explicit and unambiguous. These implications are then found to be non-trivial to reconcile with the results of said neuroimaging studies, which highlights the significance of such studies for the mind-body problem and philosophy of mind in general.

12.2 Introduction

Recently, two remarkable neuroimaging studies of the neural correlates of the psychedelic state have been completed: the first investigated the effects of psilocybin, the main psychoactive compound in magic mushrooms (Carhart-Harris et al. 2012), while the second focused on lysergic acid diethylamide, or LSD (Carhart-Harris et al. 2016). The first study has shown that, despite the significantly higher richness of experience reported by subjects on psilocybin when compared to those on placebo, measurements of Cerebral Blood Flow (CBF) with functional magnetic resonance imaging (fMRI) indicated that psilocybin caused *only reductions* of neural activity.[2] No increases in CBF

---

1 Some argue that it is more accurate to say that physicalism entails that subjective experience is entirely constituted *or generated* by brain activity, which reflects an appeal to emergence. Notice, however, that even if experience is an emergent property of brain activity, it is still *constituted* by brain activity, in the same way that e.g. the shapes of sand dunes are still constituted by sand, despite being an emergent property of sand. To say that the brain generates experience without necessarily constituting it requires *strong emergence*, as defined by Chalmers (2006). However, as I've argued elsewhere (Kastrup 2015: 59), strong emergence is incoherent: it either means nothing or entails an appeal to magic. Be it as it may, even under this allegedly more general definition of physicalism—according to which experience can somehow be generated by brain activity without being constituted by it—the argument that follows in this chapter still holds intact: there must still be a strict link between experience and brain activity in terms of information. In other words, all the information in the qualitative field of experience must still correspond to information discernible in physical arrangements. Otherwise, there would be experience ungrounded in physical dynamics: a form of disembodied phenomenality that would necessarily contradict the spirit of any ontology deserving of the label 'physicalism.'
2 A later study performed at the University of Zürich has confirmed this, showing that psilocybin causes "significantly reduced absolute perfusion" (that is, blood flow) in just about every region of the brain, whilst leading to "profound subjective drug effects" (Lewis et al. 2017).

were seen anywhere in the brain. In the second study, localized increases in CBF were observed in the visual cortex of subjects on LSD, but magnetoencephalography (MEG) — which performs a more direct measurement of neural activity than CBF — again revealed reductions in activity throughout the brain. The slight discrepancy in CBF measurements between the two studies was explained by the researchers in the following manner: "One must be cautious of proxy measures of neural activity (that lack temporal resolution), such as CBF ... lest the relationship between these measures, and the underlying neural activity they are assumed to index, be confounded by extraneous factors, such as a direct vascular action of the drug" (Carhart-Harris et al. 2016: 5). They proceeded to say that "more direct measures of neural activity (e.g. EEG and MEG) ... should be considered more reliable indices of the functional brain effects of psychedelics" (Carhart-Harris et al. 2016: 6).

The results of both studies thus indicate that the psychedelic state is consistently associated with *reductions* of brain activity, despite the significant *increases* in the richness of experience reported by subjects. From the point of view of the metaphysics of physicalism, which entails that experience is constituted by brain activity alone,[3] such results are at least counterintuitive. Indeed, neuroscientist Christof Koch commented that, *"to the great surprise of many*, psilocybin, a potent psychedelic, reduces brain activity" (Koch 2012b, emphasis added). But does this observation strictly contradict physicalism? Does physicalism imply that an increase in the richness of experience must be accompanied by an increase in brain activity?

In this brief analysis, the implications of physicalism regarding the relationship between subjective experience and brain activity will be rigorously examined from an informational perspective. The goal is to establish whether the results reported

---

3 The same earlier note about the definition of physicalism holds here.

in the neuroimaging studies cited above can be reconciled with physicalism and, if so, under what circumstances. Indeed, as neuroimaging advances and its applications begin to touch on difficult and nuanced problems in neuroscience and philosophy of mind, it becomes crucially important that the related implications of physicalism be unambiguously understood. This is what is attempted here. As such, although this brief analysis focuses only on the psychedelic studies cited, its relevance potentially extends to many more areas of neuroscientific investigation.

12.3 The implications of physicalism

Physicalism posits that there are physical entities independent of experience and that the qualities of experience are constituted by particular arrangements of such entities.[4] More specifically, under physicalism the qualities of experience are constituted by particular patterns of brain activity, which are called the 'Neural Correlates of Consciousness' (NCCs). Notice that I use the word 'activity' here in the broad and generic sense of metabolism itself, so that only a dead, non-metabolizing brain has no activity.

Not all brain activity consists of NCCs: under physicalism, there are also unconscious neural processes. Reductions in these unconscious processes don't necessarily imply reductions in experience, for they aren't NCCs. In fact, if these unconscious processes are inhibitory in nature, their reduction could even cause an increase in NCCs and, therefore, experience. As such, nothing precludes an increase in NCCs from being accompanied by a comparatively greater decrease in unconscious processes, leading to an overall decrease in brain activity. Clearly then, physicalism does *not* necessarily imply that more experience should always correlate with more *total* brain activity.

But here is the critical point: under physicalism, an increase

---

4 The same earlier note about the definition of physicalism holds here.

in the richness of experience *does* need to be accompanied by an increase in the metabolism associated *with the NCCs*, for experiences are supposedly constituted by the NCCs. Let us unpack this carefully.

Rich experiences span a broader information space in awareness than comparatively dull and monotonic experiences. This is fairly easy to see: the experience of seeing a colorful fireworks display entails more information in awareness than staring at an overcast night sky. Listening to Bach's *Brandenburg Concertos* entails more information in awareness than sitting in a relatively silent room.[5] Having an intense dream entails more information in awareness than deep sleep. And so on. There clearly are such things as *richer* and *duller* experiences.

The concept of information is crucial here: it is a measure of how many different states can be discerned in a system. *More* information means that the system comprises *more* states that can be discerned from each other (Shannon 1948). In the case of human experience, information reflects the amount of subjectively apprehended qualities that can be discerned from each other in awareness. Watching a fireworks display entails more information than staring at a dark sky because one can discern more shapes, colors, movements and levels of brightness in the former case. Listening to the *Brandenburg Concertos* entails more information than sitting in a relatively silent room because one can discern more tones, rhythms, timbres and levels of volume in the former case. To say that an experience is richer thus means that the experience entails more information in

---

5 Naturally, this assumes that all other aspects of one's inner life are equivalent in the two contrasting situations. In other words, it is assumed that the levels of e.g. thought and imagination are the same whether one is sitting in silence or listening to Bach's *Brandenburg Concertos*; whether one is staring at an overcast night sky or watching a fireworks display.

awareness.

Information states can be discerned in time (such as the progressive unfolding of notes in a symphony) and space (such as the different shapes and colors within a single snapshot of a fireworks display). In practice, however, a single moment is experientially intangible. The bulk of the information within awareness is associated with how many, and how often, qualities *change* over time. Therefore, when we speak of richer experiences we essentially mean experiences wherein a higher number of discernible qualities change more frequently.

Now, since physicalism entails that there is nothing to the qualities of experience but the states of its physical substrate,[6] an increase in the richness of experience can only be explained by more, and/or more frequent, state changes in the parts of the brain corresponding to the associated NCCs. We call these physiological state changes *metabolism*, or *neural activity*. Therefore, a relative increase in *local* metabolism is necessary to create the broader information space in the brain that supposedly constitutes the broader information space in awareness entailed by richer experiences. This is an inescapable implication of physicalism. Without it, subjective experience would become decoupled from the workings of the living brain information-wise. Operationally, thus, physicalism implies a form of *local* proportionality: the richness of experience must be proportional to the compound metabolic level *of the NCCs*, even though it doesn't need to be proportional to the *total* level of activity in the brain.

An analogy may be helpful at this point. If we model the brain as a cellular automaton (e.g. Gers, Garis & Korkin 2005), metabolism is a measure of how many, and how often, cells change states as time goes by (a 'cell' in a cellular automaton doesn't necessarily correspond to a neuron, mind you). A brain

6 The same earlier note about the definition of physicalism holds here.

displaying high activity corresponds to an automaton wherein many cells change states frequently. A brain displaying low activity corresponds to an automaton wherein a few cells change states now and then. The conclusion from the discussion above can thus be restated as follows: richer experiences, under physicalism, must correlate with an increase in the number of cells encompassed by the NCCs, and/or more frequent state changes in said cells.

Notice that this is a generic conclusion derived from first-principles informational considerations. It is independent of the exact nature of the NCCs. Neural spiking, neurotransmitter releases, fluctuations of membrane potentials, network configurations, communication or information integration patterns across neurons, etc.: whatever the NCCs turn out to be or encompass, it remains a direct implication of physicalism that *an increase in the richness of experience needs to be accompanied by an increase in the compound level of metabolism associated with the NCCs.*[7]

12.4 Interpreting the neuroimaging of the psychedelic state

Given the previous section's conclusion, what does it mean for the plausibility of physicalism that psychedelic trances are not accompanied by increases in brain activity? The first thing to consider is that psychedelic trances entail a significant

---

7 Readers of this article have pointed out that, in the psychedelic studies I cite, it is suggested that the psychedelic experience correlates with certain *patterns of synchronization* across different neural networks in the brain. But since my argument is entirely agnostic of the specific nature of the NCCs—that is, it is based solely on the notion that there must be a strict correspondence between experience and NCCs at the level of information, whatever the NCCs are—this bears no significance to my conclusions.

increase in the richness of experience when compared to an ordinary baseline. This is not only overwhelmingly attested by informal reports (such as those available online at, for instance, the 'Erowid Experience Vaults'), it has also been confirmed in controlled studies. In the first study cited above, subjects on psilocybin reported extremely vivid imagination, dream-like experiences and complex perceptual hallucinations (Carhart-Harris et al. 2012: 2138-2139), which characterized the psychedelic state unambiguously as experientially richer—i.e. spanning a broader information space in awareness—than the placebo state. In an earlier study, subjects characterized the psychedelic state as extremely rich, intense and even "more real than real" (Strassman 2001). In yet another study, 67% of the subjects rated a psychedelic experience as among the top five most spiritually significant of their life, considering "the meaningfulness of the experience to be similar, for example, to the birth of a first child or death of a parent" (Griffiths et al. 2006: 276-277). It is difficult to imagine how this could fail to imply that a psychedelic experience is richer than most other experiences in life. Thus, under physicalism, one would have expected the psychedelic neuroimaging studies cited above to have shown unambiguous local increases in brain activity corresponding to the NCCs. How can we reconcile physicalism with the fact that this was *not* the case? There are two hypotheses.

The first hypothesis is that the spatial resolution of fMRI may have been too coarse for researchers to discern between (a) hypothetical NCCs whose activity did increase and (b) unconscious processes right 'on top of' said hypothetical NCCs, whose metabolic drop masked the activity rise of the NCCs. But this possibility stretches plausibility, for it entails the rather unlikely coincidence that each and every NCC was consistently accompanied by an unconscious process intermingled with it, whose metabolism happened to decrease so significantly as to mask the corresponding NCC increase. There is no reason why these different neural processes should unfold in such a perfect spatio-temporal intermingling. Indeed, different neural processes are normally discernible from each other in neuroimaging, otherwise neuroimaging wouldn't be of much use in the first place.

The second hypothesis is that all the information entailed by the psychedelic experience—and, therefore, the corresponding level of metabolism—is *already* in the baseline brain activity of the subjects. Prior to drug intake, the information is simply not in the NCCs. In other words, the 'trip' may unfold in the brain at all times, in the form of unconscious processes. The psychedelic compound may simply convert those *existing* unconscious processes into NCCs, which then brings the trip into awareness. What this conversion may entail and how it may happen remains completely unclear and highly speculative, but the hypothesis could, at least in principle, explain why no activations were observed with respect to the placebo baseline: subjects who received placebo may have also been 'tripping' subliminally, displaying all the corresponding metabolism.

There are two problems with this second hypothesis. The first is that it implies that the brain of every person is busy physically computing a 'trip' *all the time*, below the threshold of awareness. To put this in context, notice that psychedelic 'trips' often include voyages to indescribable parallel realms; death- and birth-like experiences; conversations with what is often described as alien entities or deities; unfathomable and countless insights into the underlying nature of reality and self; the witnessing of indescribably complex structures and motion; synesthetic traversals of the entire gamut of human emotions and beyond; etc. (Strassman 2001, Strassman et al. 2008). It is at least difficult to conceive of a reason why evolution would have led to brains that systematically wasted energy and considerable cognitive resources to continuously maintain useless subliminal 'tripping.' To put it in perspective, consider for instance what can be accomplished in art or engineering with the cognitive resources associated with imagining a single complex structure in movement. Many of us have difficulties with simple 3D perspective, let alone the movements of complex structures. Yet, the hypothesis here implies that we are all subliminally wasting many more resources than this *all the time*. Such an idea seems, again, to stretch plausibility.

The second problem with the second hypothesis is this: in another brain imaging study, researchers used fMRI to measure the neural activity of subjects as they slept and dreamed (Horikawa et al. 2013). The metabolic activity corresponding to dreaming up trivial visual experiences, such as seeing a person take a photograph or staring at a bronze statue (Costandi 2013), was clearly identifiable. So the added metabolism of dreaming up trivial images is significant enough to be picked out from the baseline activity wherein, *ex hypothesi*, unfathomable psychedelic 'tripping' is continuously taking place. This suggests that the metabolic level of the hypothetical subliminal 'trip' *cannot* be overwhelmingly higher than that of the trivial dream. If it were, the activity signal of the dream would have been mere noise, indiscernible from the baseline. Yet, in terms of information richness, the experience of e.g. staring at a bronze statue is negligible in comparison to that of a full-blown psychedelic trance. Therefore, given the previous section's conclusion, the trivial dreams should have been metabolically negligible and indiscernible from the baseline, which reduces the second hypothesis to a contradiction.

In conclusion, both hypotheses conceived to reconcile physicalism with the results of recent neuroimaging studies of the psychedelic state are implausible. At present, it remains unclear if and how physicalism can accommodate such neuroimaging results. This, of course, does not mean that the results outright refute physicalism in and of themselves. Other hypotheses may exist that have not been considered in this brief analysis and further studies of the neural correlates of the psychedelic state may reframe the current results. Until more clarity is achieved, however, one is left with this sobering thought: dreams and psychedelic trances are analogous in that neither can be attributed to sensory inputs, both being entirely imagined experiences. Yet, in a dream, when one experiences something as dull as staring at a bronze statue, the corresponding brain activations can be clearly discerned by fMRI. But when one undergoes psychedelic trances rated by 67% of subjects as one of the five most significant experiences of their lives, *no conclusive activations can be discerned anywhere in the brain*.

12.5 Conclusions

The generic implications of the physicalist metaphysics regarding the relationship between the richness of experience and brain activity levels have been rigorously examined and made explicit and unambiguous. The examination was done from the perspective of informational first principles. Recent neuroimaging studies of the psychedelic state have also been reviewed and their results found to be non-trivial to reconcile with said generic implications of physicalism. This suggests that either (a) future research into the neural correlates of the psychedelic state will reframe the present results in a manner more amenable to physicalist interpretations; (b) new interpretative hypotheses will emerge to accommodate the present results under plausible physicalist scenarios; or (c) neuroimaging studies of the psychedelic state will render physicalism untenable as a metaphysical option for resolving the mind-body problem.

Part V

Related considerations

The great and pressing task of our epoch is the pure apprehension of the meaning and limits of modern science.
Karl Jaspers: *The Origin and Goal of History.*

The imagination of nature is far, far greater than the imagination of man.
Richard Feynman: *The Pleasure of Finding Things Out.*

Chapter 13

Preamble to Part V

At this point, my case for idealism has been made. Part I has shown that all other ontologies entail insoluble problems, such as the hard problem of consciousness and the subject combination problem. Moreover, these non-idealist ontologies also lack parsimony insofar as they postulate unnecessary and unprovable theoretical entities as primitives. Part II has then laid out an idealist ontology that makes sense of both classical and quantum worlds in a parsimonious manner, without falling prey to either the hard problem or the combination problem. In Part III, the best objections against idealism have been refuted, whereas Part IV has discussed empirical evidence that not only contradicts the mainstream physicalist ontology, but also directly points to idealism.

My hope is that the reader will now be at least open to idealism as a viable hypothesis, if not outright convinced of its superiority. Two very natural and legitimate questions then immediately spring forth:

(a) If physicalism is so clearly inferior to idealism, how has the former been able to dominate the mainstream cultural narrative for so long?

(b) If idealism is true, then what difference does it make regarding how best to live our lives and relate to the world?

These questions are addressed in the next two chapters. Chapter 14 shows that there have historically been strong psychological motivations for the adoption of physicalism. This may come as a surprising assertion to many, for physicalism is often perceived as a purely fact-based interpretation of reality, untarnished by subjective biases or covert wish-fulfillment maneuvers. I hope to show that little could be farther from the truth, for there are compelling reasons to believe that the physicalist worldview protects and validates the ego, even in view of formidable threats such as death. Perhaps even more surprisingly, behind physicalism's apparent denial of meaning there operate psychological mechanisms that seek, in fact, to *enhance* one's sense of meaning in life.

For these reasons, it is after all not that surprising that physicalism has come to amass the formidable level of support it has today among the intellectual elites, particularly in academia, where non-physicalist views are often considered anathema. As is often the case with views that come to define a culture, there is more to this success than just the philosophical merits — or demerits — of physicalism.

Chapter 15 then addresses the implications of idealism with respect to the significance and purpose of life in the world. Indeed, whereas physicalism denies the semantic meaning of the world by construing it to be a mechanical contraption governed by blind laws and mere chance, idealism regards the world as the symbolic appearance of what religious traditions throughout history have referred to as 'God's Mind.' This way, according to idealism, nature holds hidden but inherent semantic meaning. One could even go as far as to hypothesize that, under idealism, the telos of life is to contemplate and understand 'God's thoughts' from a perspective unavailable to 'God.' All this is carefully unpacked in Chapter 15, which represents my attempt to bring the relevance of idealism to life by highlighting how it can change the way we relate to the world.

Because it was originally published as a standalone article that thus needed to be self-contained, Chapter 15 also includes an extensive empirical argument to defend its premise that the world is entirely mental. It elaborates — in layman's terms accessible to any educated reader—upon the experimental evidence for what is technically called 'contextuality' in physics. As readers may remember, contextuality was already discussed in Chapter 6, but briefly and in rather technical terms. Chapter 15 expands on and clarifies that discussion, explaining step by step how and why several laboratory experiments over the past few decades have slowly but now surely refuted the physicalist tenet that the world does not depend on observation. As such, Chapter 15 includes what is perhaps the strongest empirical argument for idealism, so to close this book on the right note.

As a final observation, attentive readers will notice that, throughout Chapter 15, I use the word 'meaning' to denote 'sense' (as in the sense of a word or phrase), 'significance' (as in the significance of a historical moment) *and* 'purpose' (as in the purpose of an action), freely conflating all three usages. This conflation is intentional and implicitly reflects the very conclusion of the chapter: that the purpose of life is to unveil the sense and significance of the world. Thus the meaning of life in the world is simultaneously life's purpose *and* the world's sense and significance. Indeed, the very linguistic versatility of the word 'meaning' amplifies the argument in Chapter 15: 'purpose' is intrinsically connected with 'sense' and 'significance.' Perhaps language captures and preserves—like a time capsule—ancient insights we have since allowed to escape us.

Chapter 14

The physicalist worldview as neurotic ego-defense mechanism

This article first appeared in *SAGE Open*, ISSN: 2158-2440, Vol. 6, No. 4, doi: 10.1177/2158244016674515, in October 2016. *SAGE Open* is the most read journal of SAGE Publications, USA, arguably the world's premier academic publisher in the area of psychology. SAGE Publications was also the Independent Publishers Guild Academic and Professional Publisher of the Year in 2012.

14.1 Abstract

The physicalist worldview is often portrayed as a dispassionate interpretation of reality motivated purely by observable facts. In this article, ideas of both depth and social psychology are used to show that this portrayal may not be accurate. Physicalism — whether it ultimately turns out to be philosophically correct or not[1] — is hypothesized to be partly motivated by the neurotic endeavor to project onto the world attributes that help one avoid confronting unacknowledged aspects of one's own inner life. Moreover, contrary to what most people assume, physicalism creates an opportunity for the intellectual elites who develop and promote it to maintain a sense of meaning in their own lives through fluid compensation. However, because this compensatory strategy does not apply to a large segment

---

1 As I have extensively articulated earlier in this book, my position is that physicalism is demonstrably inferior to idealism on both logical and empirical grounds. Yet, since the scope of this particular article is restricted to psychology, my tone had to be neutral regarding philosophical matters.

of society, it creates a schism—with corresponding tensions—that may help explain the contemporary conflict between neo-atheism and religious belief.

14.2 Introduction

A worldview is a narrative in terms of which we relate to ourselves and reality at large. It is a kind of cultural operating system that gives us tentative answers to foundational questions such as 'What are we?' 'What is the nature of reality?' 'What is the purpose of life?' and so on (Kastrup 2014). Although many different worldviews vie for dominance today, the academically endorsed physicalist narrative defines the mainstream, despite its many difficulties (Kastrup 2014, 2015, Nagel 2012). This reigning worldview posits that physical entities outside consciousness are the building blocks of reality. Consciousness, in turn, is supposedly an epiphenomenon or emergent property of certain complex arrangements of these entities. As such, under physicalism, consciousness must be reducible to physical arrangements outside and independent of experience (Stoljar 2016).

Physicalism is often portrayed as a worldview that, in contrast to, for example, religion or spirituality, is based solely on objective facts. The present article, however, hypothesizes that the formative principles and motivations underpinning the physicalist narrative—whether it ultimately turns out to be philosophically correct or not—are partly subjective, reflecting neurotic ego-defense maneuvers meant, as described by Vaillant, to "protect the individual from painful emotions, ideas, and drives" (1992: 3). This becomes clear when one lifts core concepts of depth psychology to the social and cultural spheres. However, as a mostly clinical approach, depth psychology requires some elaboration before being applied at a theoretical level.

The modern understanding of depth psychology can be traced back to the late 19th and early 20th centuries, in the works of Frederic Myers, Pierre Janet, William James, Sigmund Freud and Carl Jung (Kelly et al. 2009). Its foundational inference is that the human psyche comprises two main subdivisions: a conscious and an 'unconscious' segment. The conscious segment of the psyche comprises experiences a person has introspective access to and can report. According to the analytical school of depth psychology, the "ego" is defined as the experiential center of this segment (von Franz 1964: 161), and it is in this specific sense that I use the word 'ego' throughout the present article. In contrast, the so-called 'unconscious' segment of the psyche comprises mental contents the person has no introspective access to and cannot report. Nonetheless, depth psychologists assert that 'unconscious' mental contents can, and do, influence the person's manifest thoughts, feelings and behaviors.

Because the ability to report an experience is a metacognitive capacity on top of the experience itself (Schooler 2002), a more rigorous articulation of the difference between the conscious and 'unconscious' segments of the psyche is this: conscious mental contents are those a person *both* experiences *and* knows that he or she experiences them. 'Unconscious' mental contents, on the other hand, are those the person either does not experience or does not know *that* he or she experiences them (Kastrup 2014: 104-110). In other words, conscious mental contents fall within the field of egoic self-reflection and, therefore, can be reported, whereas 'unconscious' mental contents escape this field and, therefore, cannot be reported. Indeed, the existence of mental contents that are experienced but cannot be reported—even to oneself—is now well established in neuroscience, which has prompted the emergence of so-called "no-report paradigms" (Tsuchiya et al. 2015).

However, as clinical psychologists can only gauge consciousness based on what their patients report, anything outside the field of self-reflection is indistinguishable from true unconsciousness. This explains the somewhat inaccurate terminology choice of the founders of depth psychology.[2]

Some critics have questioned the existence of an 'unconscious' segment of the psyche on philosophical grounds (Stannard 1980: 51-81). However, recent empirical results in neuroscience show the presence of broad cognitive activity that individuals cannot report, but which nonetheless causally conditions the individuals' manifest thoughts, feelings or behaviors (Augusto 2010, Eagleman 2011, Westen 1999). Recent neuroimaging studies of the psychedelic state have also corroborated the depth-psychological view that ego suppression—in the form of reduction of neural activity in the brain's default mode network—brings otherwise 'unconscious' mental contents into awareness (Carhart-Harris et al. 2012, Carhart-Harris et al. 2016, Palhano-Fontes et al. 2015).

On the basis of these empirical results, the core idea of depth psychology—that is, that a segment of the psyche that escapes self-reflective introspection can causally condition our thoughts, feelings and behaviors—cannot be dismissed. And because cultural narratives are the compound result of an aggregation of the thoughts, feelings and behaviors of individuals, depth-psychological insights are valid starting points for an analysis of the psychological underpinnings of our culture's mainstream worldview.

In Sections 14.3 and 14.4, I review ways in which the physicalist narrative can give us permission to avoid confronting unwanted affects in the 'unconscious' segment of our psyche. In Section 14.5, I elaborate on how physicalism can conceivably

---

2 See Chapter 9 for a much more extensive elaboration on the nature of the 'unconscious,' including the role of dissociation, which I have not discussed in this particular article. In a nutshell, my position is that there is no actual unconscious, but simply *conscious* mental processes inaccessible to egoic introspection because they (a) escape the field of self-reflection or (b) are strongly dissociated from the ego.

even nurture its proponents' sense of meaning in life. This latter section is based on theories of social psychology, rather than depth psychology, but it still leverages the notion of an 'unconscious': in hypothesizing that physicalism is an expression of fluid compensation, it presupposes that cognitive processes outside the field of self-reflection influence the feelings, thoughts and opinions subjects express. Finally, Section 14.6 briefly sums up the key ideas defended in this article.

14.3 Ego protection through projection

According to depth psychology, a neurosis is the expression of an inner psychic conflict caused by the ego's refusal to acknowledge, confront and ultimately integrate unwanted affects rising from the 'unconscious' (Jung 2014: 137). To keep these affects at bay, the ego uses a variety of defense mechanisms, among which denial, distortion, dissociation, repression and so on (Vaillant 1992). A particularly common defense mechanism is *projection* (Ibid.), whereby one circumvents the need to confront ego-threatening forces within oneself by ascribing the corresponding attributes to the outer environment. As such, projections can be said to partly hijack and manipulate one's worldview in an attempt to prevent short-term suffering. My hypothesis is that, through projection, the physicalist worldview gives us permission to avoid confronting some of what we find disagreeable within ourselves. This can be achieved in a variety of subtle ways.

For instance, we all have a sense of our own existence and identity. Lucid introspection reveals that the root of this sense is our consciousness—our capacity to be subjects of experience. After all, if we were not conscious, what could we know of ourselves? How could we even assert our own existence? Being conscious is what it means to *be* us. In an important sense— perhaps even the *only* important sense—we are first and foremost consciousness itself, the rest of our self-image arising afterward, as thoughts and images constructed *in* consciousness.

From this perspective, the physicalist narrative's attempt to reduce consciousness to physical entities outside subjectivity is counterintuitive, for it divorces the alleged nature of consciousness from our felt sense of identity. We do not *feel* as though we were a bunch of physical particles bouncing around inside our skull. Instead, we feel that we are the subjective 'space' wherein our experiences unfold, including our ideas about physical particles. Hence, there is a sense in which the physicalist narrative can be said to *project* the felt essence of ourselves onto something distinctly other. According to it, we are not really 'here,' grounded in our subjective sense of being, but somewhere 'over there,' in an abstract world fundamentally beyond the felt concreteness of our inner lives. As such, the physicalist narrative entails an *emptying out* of what it means to be us; a kind of secular kenosis. "I am no ghost, just a shell," laments the art character Annlee (Huyghe & Parreno 2003: 35), whose predicament is that of many of us in contemporary society.

The kenosis entailed by the physicalist narrative can exonerate its proponents from responsibility for their choices and actions. Consider this passage by Sam Harris: "Did I consciously choose coffee over tea? No. *The choice was made for me by events in my brain* that I ... could not inspect or influence" (2012a: 7-8, emphasis added). The projection of responsibility here is clear and the corresponding release described by Harris himself: "Losing a belief in free will has not made me fatalistic—in fact, *it has increased my feelings of freedom*. My hopes, fears, and neuroses seem *less personal*" (2012a: 46, emphasis added). Indeed, under the ethos of such a worldview, there is no concrete reason for guilt or regret, for we allegedly are not what we experience ourselves to be. We are not responsible for what happens *here* because we are not—and have never been—really *here*. We are not ghosts in the machine but ghosts *conjured up by* the machine. In a significant sense, we do not really exist.

As a matter of fact, some proponents of the physicalist narrative go as far as to deny that consciousness exists. "Consciousness doesn't happen. It's a mistaken construct." These words of neuroscientist Michael Graziano (2016) should give anyone pause for thought. Here we have consciousness — whatever it may intrinsically be — denying that consciousness exists. Philosopher Daniel Dennett (1991) also claimed that consciousness is an illusion, a claim that seems to immediately contradict itself. After all, where do illusions occur if not in consciousness?[3] By appealing to metaphysical abstractions fundamentally beyond experience, such denials of our felt selves achieve a form of deliverance somewhat analogous to religious absolution. Surprisingly, as we will later see, they even help restore a sense of meaningfulness in life, following what I will call 'ontological trauma.'

The structure of these denials is fairly clear: first, consciousness weaves the conceptual notion that certain aspects of its own dynamics somehow exist outside itself; then, it projects its own essence onto these aspects. The corresponding dislocation of identity is apparent — and its neurotic character easy to grasp — with an analogy: imagine a painter who, having painted a self-

---

3 In the words of David Bentley Hart, "The entire notion of consciousness as an illusion is, of course, rather silly. Dennett has been making the argument for most of his career, and it is just abrasively counter-intuitive enough to create the strong suspicion in many that it must be more philosophically cogent than it seems, because surely no one would say such a thing if there were not some subtle and penetrating truth hidden behind its apparent absurdity. But there is none. The simple truth of the matter is that Dennett is a fanatic: He believes so fiercely in the unique authority and absolutely comprehensive competency of the third-person scientific perspective that he is willing to deny not only the analytic authority, but also the actual existence, of the first-person vantage" (2017).

portrait, points at it and declares himself to *be* the portrait. This, in essence, is what physicalists do, whether it is philosophically justifiable or not.[4] Their consciousness conceptualizes self-portraits within itself. Sometimes these self-portraits take the form of electrical impulses and neurotransmitter releases in the brain (Koch 2004). Other times, they take the shape of quantum transitions or potentials (Tarlaci & Pregnolato 2016). Whatever the case, their consciousness always points to a conceptual entity it creates within itself and then declares itself to *be* this entity. It dismisses its own primary, first-person point of view in favor of an abstract third-person perspective. Consider Dennett's words: "The way to answer these 'first-person point of view' stumpers is *to ignore the first-person point of view* and examine what can be learned from the third-person point of view" (1991: 336, emphasis added). The contempt for the *subject* of experience — the primary datum of existence and one's own felt identity — is palpable here; the kenosis nearly total.

The physicalist narrative may also give us permission to carve out and dismiss — again through the kenosis of projection — the most difficult aspects of our inner lives: our felt emotions. According to it, the feeling of an emotion is the internal perception of an "action program" triggered by certain stimuli (Damasio 2011). Although the action program itself is important insofar as it helps us survive and reproduce, the accompanying feeling of emotion is, in a sense, a mere side effect of the program's execution. For instance, the sight of another human being facing a predicament is a stimulus that triggers actions meant to help the victim and, consequently, increase the social cachet of the action taker. The *feeling* of compassion, in turn, is supposedly nothing but the inner perception of this evolutionarily useful

---

4 For clarity, and at the cost of repeating myself, my position is that this is *not* philosophically justifiable, as I have extensively argued earlier in this book.

reactive schema (Immordino-Yang et al. 2009); it allegedly has no primary or fundamental significance. Under such a narrative, it is easier to go into denial about our emotional lives when the going gets tough. We feel justified to dismiss or repress our traumas and demons, avoiding the often-painful work of psychological integration. The physicalist narrative provides a foundation for rationalizing the choice of living an unexamined, superficial life. To a person desperate to avoid the specter of immediate and pungent suffering, the benefits of this stance may seem to far outweigh its potential long-term implications.

Surprisingly, the physicalist narrative can even offer us reassurance about death. According to it, there is literally nothing to fear about death itself, because it is allegedly the end of all experience, including the experiences of fear and pain. All of our problems and suffering are guaranteed to end at that point. The great and scary *unknown* of the experiential realm beyond physical existence vanishes in one fell swoop; the greatest angst of humankind is conquered. The psychological allure of this idea is powerful, yet most people do not seem to ever stop to consider it. We have come to take for granted the comforts that our mainstream worldview grants us.

To sum it up, by denying our felt sense of existence and identity, the physicalist narrative creates an opportunity to clear the ego of ultimate responsibility. By denying the fundamental reality of emotions, it creates an opportunity to protect the ego from a confrontation with far more powerful forces. And by projecting our ontological essence onto ephemeral arrangements of matter, it creates an opportunity to protect the ego from what has historically been the greatest angst of humankind: the experiential unknown of the after-death state.

14.4 Egoic control

It has been shown that religiosity can reflect a form of compensatory control (Kay et al. 2010): by believing that transcendent forces aligned with one's convictions govern the world, the ego avoids the anxiety associated with its own inability to overcome uncertainty. This way, religiosity creates an opportunity for control *by proxy*: although the ego cannot determine the course of nature, an external agency far superior to it is believed to do so in a way consistent with the ego's preferences. The ego's need to avoid anxiety by exerting control is thus *indirectly* fulfilled.

Going beyond religiosity, the physicalist narrative enables a sense of *direct* egoic control over nature. Indeed, a recent empirical study has shown that "believing that science is or will prospectively grant ... mastery of nature imbues individuals with the belief that they are in control of their lives" (Stavrova, Ehlebracht & Fetchenhauer 2016: 234). Of course, by associating itself with science—in a philosophically questionable move that is nonetheless widely accepted—the physicalist narrative has become the enabler and ontological foundation of this belief. And because *direct* control—the notion that one can personally steer or at least predict what is going to happen—is known to be a key contributor to mental well-being (Langer & Rodin 1976, Luck et al. 1999), it stands to reason that the allure of physicalism in this regard could potentially be even stronger than that of religious control-by-proxy.

The opportunity for direct control offered by the physicalist narrative goes as far as conquering death itself: if consciousness is just an epiphenomenon or emergent property of physical arrangements outside experience, it becomes conceivable that, through smart engineering, we could create means to upload our consciousness into more durable substrates such as silicon computers (Kurzweil 2005). Some physicalists even offer detailed roadmaps for achieving this (Sandberg & Boström 2008). The possibility of eternal life thus seems to open up, provided that consciousness can be instantiated in a computer by programming the computer with the patterns of information flow found in a person's brain.

This, however, is premised on the notion that a simulation of a mental phenomenon is equivalent, *in essence*, to the phenomenon itself. There are many compelling arguments against this notion in philosophy of mind, the most well known of which is perhaps John Searle's (2004). To gain some intuition about what these arguments generally entail, consider this: Do we have any reason to believe that, by performing a perfectly accurate simulation of kidney function in a computer, the computer will begin urinating on its desk? Clearly not. There is an essential difference between a computer simulation and the phenomenon it simulates; they are not the same thing, no matter how accurate the simulation.[5] Yet, those hoping to 'upload consciousness' under the physicalist narrative seem to become so engrossed in abstraction that they lose touch with basic intuitions of plausibility. Their neurosis is, in this sense, comparable with religious dogmatism.

Although both the religious and physicalist narratives create an opportunity for conquering death, the Promethean door to immortality opened by physicalism invests the *ego*— not deities—with the power to control transcendence through technology. This is seductively more direct, its only weakness— from a psychological standpoint—being that it is promissory: at present, nobody has ever managed to upload consciousness. Yet, some popular physicalist authors argue that consciousness uploading may be achievable *still in our own lifetime* (Kurzweil 2005, Sandberg & Boström 2008), which actualizes the potential allure of their worldview.

As seen in Section 14.3 and this section, the implications of the physicalist narrative consistently help protect and invest the ego with authority. This is not to say that physicalism is entirely motivated by neurotic ego-defense maneuvers, for

---

5 A much more extensive elaboration of this point can be found in Chapter 5, Section 5.10.

there is a philosophical argument behind it that cannot be dismissed.[6] Nonetheless, the question is whether it is plausible that physicalism's significant ego-defense potential has *not* been, to some degree, an unexamined motivation for its development, promotion and adoption.

14.5 The question of meaning

Meaning—in the sense of significance and purpose—is probably the greatest asset any human being can possess. Psychotherapist Victor Frankl (1991), who practiced and led groups while detained in a concentration camp during World War II, asserted that the *will-to-meaning* is the most dominant human drive, in contrast to Nietzsche's will-to-power and Freud's will-to-pleasure. Meaning is so powerful that, as Jung remarked, it "makes a great many things endurable—perhaps everything" (1995: 373). Philip K. Dick's alter ego Horselover Fat, in the novel *Valis*, embodies the essence of this drive: "Fat had no concept of enjoyment; he understood only meaning," wrote Dick (2001: 92). Like Fat, many of us see meaning as a higher value than power or pleasure. Our motivation to live rests in there being meaning in our lives. Today, we need meaning more than ever, for as Paul Tillich (1952) lucidly observed, the greatest anxieties of our culture are precisely those of *doubt* and *meaninglessness*.

And here is where an argument is often made for the impartiality of physicalism: as a worldview that, by turning the universe into a mechanical contraption fueled by mere chance, drains the meaning out of life, it cannot possibly be a neurotic ego-defense mechanism—or so the argument goes. Instead, the physicalist narrative must represent a courageous admission by "tough people who face the bleak facts" (Watts 1989: 65). It must embody an objective assessment of reality, not an

---

6 It cannot be dismissed but it *can* be refuted on both logical and empirical grounds, as I hope to have done in Parts I to IV of this book.

emotional, irrational wish-fulfillment maneuver akin to religion. Compelling as it may seem at first, this argument fails careful scrutiny, for its premise is false.

Indeed, according to the Meaning Maintenance Model (MMM) of social psychology (Heine, Proulx & Vohs 2006) — which is perhaps better seen in the context of a broader theory of psychological defense (Hart 2013) — we can derive a sense of meaning from four different sources: self-esteem, closure, belonging, and symbolic immortality. In other words, we can find meaning in life through (a) cultivating a feeling of personal worth, (b) resolving doubts and ambiguities, (c) being part of something bigger and longer-lasting than ourselves, and (d) leaving something of significance behind — such as professional achievements — in the form of which we can 'live on' after physical death. A society's mainstream cultural narrative conditions how meaning can be derived from each of these four sources.

The key idea behind the MMM is that of *fluid compensation* as an ego-defense mechanism: If one of the four sources of meaning is threatened, an individual will tend to automatically compensate by seeking extra meaning from the other three sources. For instance, threats to self-esteem may cause the individual to reaffirm his or her model of reality, thereby bolstering closure.

As van Tongeren and Green (2010) have shown, a transcendent source of meaning, such as religion, plays the same role in fluid compensation as the other four sources. For instance, individuals tend to reaffirm their religious beliefs following disruption to their meaning system, in an effort to protect the latter. Van Tongeren's and Green's experiments have not only empirically substantiated the MMM, they have also shown that even *subliminal* threats to meaning trigger fluid compensation, strongly indicating that the 'unconscious' is integral to the process.

With this as background, my suggestion is that the physicalist narrative, in addition to being a rational hypothesis for making sense of the world,[7] may be an expression of fluid compensation by intellectual elites. In other words, instead of a threat to meaning, the physicalist narrative may actually reflect an attempt by these elites to protect and restore their sense of meaning through bolstering closure, self-esteem and symbolic immortality. The disruption that may have originally led to this compensatory move occurred around the mid- to late-19th century.

Indeed, it was at that time that we lost our ability to spontaneously relate to religious myths without linear intellectual scrutiny. "With Descartes and Kant, the philosophical relation between Christian belief and human rationality had grown ever more attenuated. By the late nineteenth century, with few exceptions, that relation was effectively absent," wrote Tarnas (2010: 311). The myths that had hitherto offered us meaning through the promise of *literal* immortality and metaphysical teleology became untenable. Taylor, who richly chronicled this historical transition, characterized the corresponding loss of meaning rather broadly and generally as "a wide sense of malaise at the disenchanted world, a sense of it as flat, empty" (2007: 302). He even hinted at fluid compensation when speaking of "a multiform search for something within, or beyond [the world], which could compensate for the meaning lost with transcendence" (Ibid.).

While acknowledging that this generalized malaise was the matrix of what followed, I submit that a more specific, forceful and *personal* threat to meaning was necessary to mobilize the extraordinary level of academic and intellectual endorsement amassed by physicalism. After all—as Taylor himself described through what he called "the nova effect"—the malaise, in and

---

7 Again, here I am making a charitable concession to physicalism because the limited scope of this particular article—focused, as it is, on psychology—prevented me from arguing against it philosophically.

by itself, fostered not only physicalism but also an explosion of myriad other worldviews.

I hypothesize that a profound and disturbing change in the intellectual elites' understanding of the nature *of their own being* — that is, an ontological trauma — was the specific, forceful and personal trigger that helped congeal the physicalist narrative. Having lost religion, the elites were left with the prospect of physical deterioration without the path to transcendence previously offered by an immortal soul. Hence, they were forced to face the inexorability of their own approaching death. And as we know from Terror Management Theory, mortality salience is a formidable threat to meaning (Pyszczynski, Greenberg & Solomon 1997) empirically shown to motivate investment in palliative worldviews (Burke, Martens & Faucher 2010). Ontological trauma may have thus triggered fluid compensation and ultimately led to the intellectual elites' championing of the physicalist narrative.

Indeed, many studies have shown that mortality salience leads to a heightened need for *closure* (Landau et al. 2004). This is fluid compensation in action. Notice also that the physicalist narrative is humanity's most significant attempt yet to achieve *closure* in our worldview. As multibillion-dollar experiments like the Large Hadron Collider — whose primary purpose is to 'close' the Standard Model of particle physics, with no immediate practical applications — illustrate, physicalism embodies an unprecedented effort to produce a causally complete, unambiguous model of reality. Nothing else in millennia of preceding history has come anywhere near it. I suggest that this is not coincidental: the physicalist narrative may reflect the elites' ego's attempt to regain, through heightened *closure*, the meaning it lost along with religion. Moreover, other modalities of fluid compensation may be at play here as well: by distinguishing themselves as a segment of society uniquely capable to understand facts and concepts beyond the cognitive capacity of others, the scientists and academics who promote the physicalist narrative stand to gain in self-esteem. The cosmological scope of the scientific work they produce and leave behind upon their deaths can also be seen as a boost to symbolic immortality. Finally, recall Tillich's observation: *doubt* and *meaninglessness* anxiety dominate our culture's mindset. Is it humanly plausible that our mainstream narrative would have evolved to tackle only doubt and leave meaninglessness anxiety unaddressed?

All in all, the physicalist narrative does not necessarily represent a net loss of meaning for the intellectual elites who produced and continue to promote it. The transcendent meaning lost along with religion may be compensated for by an increase in closure, self-esteem and symbolic immortality. Unfortunately, however, this compensatory strategy cannot work for most ordinary people: the men and women on the streets do not have enough grasp of contemporary scientific theories to experience an increase in their sense of closure. Neither do they gain in self-esteem, because they are not part of the distinguished elites. Finally, insofar as ordinary people do not produce scientific work of their own, no particular gain in symbolic immortality is to be expected either.

In conclusion, the physicalist narrative may serve the egoic meaning needs of the intellectual elites who develop and promote it, but constitutes a significant threat to the sense of meaning of the average person on the streets. Perhaps for this reason, a large segment of society seeks meaning through alternative ontologies considered outdated and untenable by the intellectual elites, such as religious dualism (Heflick et al. 2015). This creates a schism—with corresponding tensions—between different segments of society, which may help explain the contemporary conflict between neo-atheism and religious belief.

14.6 Conclusion

The physicalist narrative, in contrast to the way it is normally portrayed, may not be dispassionate. It may be partly driven by the neurotic endeavor to project onto the world attributes that help us avoid confronting unacknowledged aspects of our own inner lives. Moreover, contrary to what most people assume, physicalism creates an opportunity for the intellectual elites who develop and promote it to maintain a sense of meaning in their own lives through fluid compensation. However, because this compensatory strategy does not apply to a large segment of society, it creates a schism—with corresponding tensions—that may help explain the contemporary conflict between neo-atheism and religious belief.

Chapter 15

Not its own meaning: A hermeneutic of the world

This article first appeared in *Humanities*, ISSN: 2076-0787, Vol. 6, No. 3, Article No. 55, on 2 August 2017. *Humanities* is published by MDPI AG, Basel, Switzerland. According to the website openaccess.nl, sponsored by Dutch universities and research institutes to foment the publication of publicly-funded research in open-access journals, MDPI AG was one of the most popular open-access publishers amongst Dutch academics in 2016.[1]

15.1 Abstract

The contemporary cultural mindset posits that the world has no intrinsic semantic value. The meaning we see in it is supposedly projected onto the world by ourselves. Underpinning this view is the mainstream physicalist ontology, according to which mind is an emergent property or epiphenomenon of brains. As such, since the world beyond brains isn't mental, it cannot *a priori* evoke anything beyond itself. But a consistent series of recent experimental results suggests strongly that the world may in fact be mental in nature, a hypothesis openly discussed in the field of foundations of physics. In this essay, these experimental results are reviewed and their hermeneutic implications discussed. If the world is mental, it points to something beyond its face-value appearances and is amenable to interpretation, just as ordinary dreams. In this case, the project of a Hermeneutic of Everything

---

1 See: http://openaccess.nl/en/what-is-open-access/open-access-pub-lishers (accessed on 24 April 2017). The Netherlands is my home country and the location of my *alma mater*.

is metaphysically justifiable.

15.2 Introduction

To be amenable to interpretation, things and phenomena must point beyond themselves, thereby embodying semantic value or sense. For instance, these squiggles of ink on paper—which we call written words—mean more than just squiggles of ink on paper: they point to something beyond themselves. Similarly, the inner imagery we experience in dreams points to something beyond their face-value appearances, which has motivated depth psychologists to develop extensive hermeneutics of dreams (e.g. Ackroyd 1993, von Franz & Boa 1994, Jung 2002, Fonagy et al. 2012). Finally, the symbolisms of religious myths point to something that transcends the face-value appearances of the symbols themselves and engages people at an emotional level (Kastrup 2016a).

Influenced by twentieth century positivism and existentialism, the contemporary cultural mindset posits that things and phenomena only have semantic value insofar as we project this value onto them. Summarizing the essence of this mindset, Sartre wrote: "there exist concretely alarm clocks ... But ... then I discover myself suddenly as the one who gives its meaning to the alarm clock ... the one who finally makes the values exist" (1992: 77). Analogously, squiggles of ink mean more than squiggles of ink only insofar as we stipulate by convention that they do so. To the extent that alarm clocks and written words are inventions of human beings, it is reasonable to assert that their meaning consists in what we project onto them.

However, the contemporary cultural mindset extends this notion of projected meaning to nature itself. Fire only represents "the inseminating fury of sex and the ardor of the ascetic" (Ronnenberg & Martin 2010: 84) insofar as we project passion onto it. Stones only represent eternity (Ibid.: 106) insofar as we project timelessness onto them. Without our projections, stones mean just stones; fire means just fire. In and of itself, the world supposedly is its own meaning. It does not inherently point to anything beyond its own appearance on the screen of perception. Whatever sense we may see in a fact of the world is supposedly a confabulation of human cognition, not intrinsic to the fact itself. "In this case," as Zemach put it, "one may say either that this fact has no sense, or that the only sense it has is provided by its form" (2006: 363). In other words, "The sense of the world is identical with its form" (Ibid.: 367). Ortiz-Osés put it perhaps most simply: "When taken 'existentially,' existence seems to lack sense, whereas sense taken 'essentially' would appear to lack existence" (2008: 65).

As a result, our culture believes that the semantic value of the world is simply an artifact of human minds. The world doesn't have a story to tell, a suggestion to make or an insight to convey. It isn't saying anything. There is nothing meaningful to be gleaned from the world, just utilitarian predictions to be made about its behavior. Under such ethos, projects such as Ortiz-Osés' —meant to formulate a symbolic hermeneutic of the world premised on the notion that "the whole of existence contains an almost secret essence" (2008: 1)—become metaphysically precarious, which Ortiz-Osés himself seems to have acknowledged (Ibid.: 65).

At the root of this state of affairs is the split between mind and world that characterizes our present worldview. Indeed, according to the mainstream physicalist ontology, the fundamental building blocks of reality are physical elements that exist independently of mind (Stoljar 2016). The latter, in turn, is supposedly constituted or generated by particular local arrangements of these physical elements, such as brains inside skulls. Consequently, mind is insulated from the external world surrounding it beyond the skull.

The problem, of course, is that only mind can host intrinsic semantic value, for the latter consists of cognitive associations: the intrinsic meaning of an experience is the emotions, insights and inner imagery it evokes. For instance, the feeling of hunger may evoke inner imagery related to food because there is a cognitive association between the feeling and the imagery. A memory from childhood may evoke the emotion of happiness because there is a cognitive association between the memory and the emotion. These associative links are an exclusive feature of mentation.

So if semantic value is essentially mental and mind is insulated from the world beyond the skull, then semantic value cannot exist in the world. A non-mental world can *be* evoked, but it cannot intrinsically *evoke* anything. Such separation between meaning and world is what motivates our contemporary culture to consider the world semantically mute. "The human mind has abstracted from the whole all ... meaning, and claimed [it] exclusively for itself," wrote Tarnas (2010: 432).

Within mind, cognitive associations can go on indefinitely, as endless chains of evocations: a daydream may lead to a thought, which may evoke an emotion, which may trigger a memory, which may lead to another thought, and so on (Karunamuni 2015: 2-3). But once we leave the inner space of mentation by evoking an external fact in the world, the chain must end. The world is the chain's final destination, for it cannot *a priori* evoke anything else in turn. This semantic end point is what we call a 'literal fact.' Everything prior to it is sign, simile or allegory—roundabout, indirect ways to arrive at the destination. According to our contemporary cultural mindset, the value of these indirections is entirely conditioned upon their ability to ultimately point at literal facts. Anything short of it is considered delusion, for it allegedly can't be anchored in truth.

But does our current scientific understanding of reality truly corroborate this split between mind and world, inside and outside? Are we justified in taking for granted that the world 'out there' is fundamentally distinct or separate from the mind 'in here'? If not, could the world carry intrinsic semantic value and be amenable to interpretation, just as dreams are? Could there be a valid hermeneutic of the world, a vision of it as symbolic, suggestive of something beyond its own face-value appearances on the screen of perception? What would the implications of this possibility be for the way we relate to the world? These are the questions addressed in this essay.

In Section 15.3, the latest experimental results emerging from the field of quantum physics will be briefly reviewed. They empirically indicate that mind and world aren't, after all, fundamentally distinct or separate. Section 15.4 will show how this continuity between mind and world can explain why the axioms of rational thought describe and model the world so uncannily accurately. In Section 15.5, the hermeneutic implications of the mental world hypothesis will be discussed. Section 15.6 then compares the analysis in Section 15.5 with what some of the world's philosophical and spiritual traditions have to say about the nature and meaning of the world. Finally, Section 15.7 concludes this essay with a brief discussion.

15.3 The ontological status of the world

The mainstream physicalist notion that the world is outside and independent of mind is an abstract explanatory model constructed in thought, not an empirical observation. After all, what we call 'the world' is available to us solely as 'images'—defined here broadly, so to include any sensory modality—on the screen of perception, which is itself mental. We *interpret* the contents of perception as coming from a world outside mind because this seems to explain the fact that we all share the same world beyond the boundary of our skin, as well as the fact that the laws that govern this world do not depend on our personal volition. Stanford physicist Prof. Andrei Linde, well known for his theories of cosmological inflation, summarized it thus:

Let us remember that our knowledge of the world begins not

with matter but with perceptions. I know for sure that my pain exists, my "green" exists, and my "sweet" exists. I do not need any proof of their existence, because these events are a part of me; everything else is a theory. Later we find out that our perceptions obey some laws, which can be most conveniently formulated if we assume that there is some underlying reality beyond our perceptions. This model of material world obeying laws of physics is so successful that soon we forget about our starting point and say that matter is the only reality, and perceptions are only helpful for its description. This assumption is almost as natural (and maybe as false) as our previous assumption that space is only a mathematical tool for the description of matter. But in fact we are substituting reality of our feelings by a successfully working theory of an independently existing material world. And the theory is so successful that we almost never think about its limitations until we must address some really deep issues, which do not fit into our model of reality. (1998: 12)

This model of reality has intuitive implications amenable to confirmation—or refutation—through subtle experimental arrangements, which Linde alluded to when he spoke of "some really deep issues." Indeed, the properties of a physicalist world should exist and have definite values even when this world is not being observed: the moon should exist and have whatever weight, shape, size and color it has even if nobody is looking at it. Moreover, a mere act of observation should not change the values of these properties: the weight, shape, size and color of the moon should not become different simply because someone happened to look at it.

Operationally, these intuitive tenets of physicalism are translated into the notion of 'non-contextuality': the outcome of an observation should not depend on the way other, separate but simultaneous observations are performed. After all, the properties being observed are supposed to be independent of observation. What I perceive when I look at the night sky should not depend on the way other people look at the night sky along with me, for the properties of the night sky uncovered by my observation should not depend on theirs. Clearly—and in line with physicalism—non-contextuality implies that the world is independent of perception, insofar as perception constitutes observation. My perceptions should simply *reveal* what the properties of the world are in and of themselves.

The problem is that, according to quantum theory, the outcome of an observation *can* depend on the way another, separate but simultaneous observation is performed. For instance, if two particles A and B are prepared in a special way, the properties of particle A as seen by a first observer— say, Alice—are predicted to correlate with the way another observer—say, Bob—simultaneously looks at particle B. This is so even when A and B—and, therefore, Alice and Bob—are separated by arbitrarily long distances. For instance, what Alice sees when she looks at particle A in, say, London, depends on the way Bob concurrently looks at particle B in, say, Sydney. If the properties of the world were outside and independent of Alice's and Bob's minds—that is, outside and independent of their perceptions—this clearly shouldn't be the case; unless there is some observation-independent hidden property, covertly shared by A and B *and entirely missed by quantum theory,* which could account for the correlations. This was Einstein's point when he (in)famously suggested that quantum theory was incomplete (Einstein, Podolsky & Rosen 1935). However, as mathematically proven by John Bell (1964), the correlations predicted by quantum theory cannot be accounted for by these kinds of observation-independent hidden properties.

Consequently, quantum theory appears to contradict non-contextuality and render physicalism untenable. A conceivable way to avoid this conclusion while accepting quantum theory would be to posit that particles A and B, or Alice and Bob themselves, somehow 'tip each other off' during observation, *instantaneously and at a distance,* so to coordinate their actions and produce the predicted correlations. This, however, would require faster-than-light communication and fly in the face of the overwhelmingly confirmed theory of special relativity.

Alternatively, a physicalist could attempt to salvage non-contextuality and the notion of a world outside and independent of mind by rejecting quantum theory itself. Yet, as it turns out, since Alain Aspect's seminal experiments (Aspect, Grangier & Roger 1981, Aspect, Dalibard & Roger 1982, Aspect, Grangier & Roger 1982)[2] the predictions of quantum theory in this regard have been repeatedly confirmed, with ever-increasing rigor. For instance, in an experiment performed in Geneva, Switzerland, in 1998 (Tittel et al.), the particles A and B were separated by more than 10 km—as opposed to the 12 meters of Aspect's original experiment (1981)—reducing the already low likelihood that they could be creating the correlations predicted by quantum theory through some kind of signal exchange. Despite this greater separation, the predictions of quantum theory were again confirmed.

Then, still in 1998 but this time in Innsbruck, Austria, another experiment (Weihs et al.) was done to eliminate another far-fetched possibility: that, *in advance* of the preparation of particles A and B, 'Alice,' 'Bob' and the system responsible for the preparation could somehow be 'pre-agreeing' on a hidden plan of action, so to later create the correlations without need for faster-than-light communication ('Alice' and 'Bob,' in this case, were automated measurement apparatuses). To close this

---

2 According to Kafatos and Nadeau, the results of these experiments "represented another signal event in a revolution in thought that was extending itself well beyond the narrow concerns of quantum physicists or philosophers and historians of science" (1990: 2).

unlikely 'conspiracy' loophole, the behaviors of 'Alice' and 'Bob' were programmed randomly and only *after* particles A and B had already been prepared. Nonetheless, the correlations predicted by quantum theory were yet again confirmed.

Critics continued to speculate about other far-fetched loopholes in these experiments. In an effort to address and close all conceivable loopholes, Dutch researchers have recently performed an even more tightly controlled test, which—unsurprisingly by now—echoed the earlier results (Hensen et al. 2015). This latter effort was considered by the periodical *Nature* the "toughest test yet" (Merali 2015). Given all this, it seems now untenable to argue against the veracity of quantum theory.

The only alternative left for physicalists is to try to circumvent the need for faster-than-light signal exchanges by imagining and postulating some form of non-locality: nature must have—or so they speculate—observation-independent hidden properties that are *not* confined to particular regions of spacetime, such as particles A and B. In other words, the argument is that the observation-independent hidden properties allegedly missed by quantum theory are 'smeared out' across space and time. It is this omnipresent, invisible but objective background that supposedly orchestrates the correlations predicted by quantum mechanics. Non-contextuality and physicalism can thus be salvaged; or can they?

The problem, of course, is that non-local hidden properties are arbitrary: they produce no predictions beyond those already made by standard quantum theory. As such, it could be argued that they represent an effort "to modify quantum mechanics to make it consistent with [one's] view of the world," so to avoid the need "to modify [one's] view of the world to make it consistent with quantum mechanics" (Rovelli 2008: 16).

Be it as it may, it turns out that certain specific correlations predicted by quantum theory are incompatible with non-contextuality *even for large classes of non-local hidden properties*

(Leggett 2003). Studies have now experimentally confirmed these correlations (Gröblacher et al. 2007, Romero et al. 2010), thus putting non-contextuality in even more serious jeopardy. To reconcile these results with physicalism would require a profoundly counterintuitive redefinition of what we call 'objectivity.' And since our contemporary cultural mindset has come to associate objectivity with reality itself, the science press felt compelled to report on some of these results by pronouncing, "Quantum physics says goodbye to reality" (Cartwright 2007).

More recent experiments have again contradicted non-contextuality and confirmed that, unlike what one would expect if the world were separate or distinct from mind, the observed properties of the world indeed cannot be said to exist prior to being observed (Lapkiewicz et al. 2011, Manning et al. 2015). For all intents and purposes, the world we perceive is *a product of observation*. Commenting on this, physicist Anton Zeilinger has been quoted as saying that "there is no sense in assuming that what we do not measure [that is, observe] about a system has [an independent] reality" (Ananthaswamy 2011).

So the question now is: Can some form of physicalism survive the failure of non-contextuality? We have seen earlier that the intuitive tenets of physicalism are: (a) there exists a world outside mind; and (b) mere observation doesn't change this independently existing world. The failure of non-contextuality clearly rules out (b). Can (a) still make any sense in the absence of (b)? If it can, then the world outside mind must somehow *physically change, instantaneously*, every time it is observed. The plausibility of this notion aside, notice that one never gets to see the observation-independent world, for it supposedly changes instantly, in an *observation-dependent* manner, the moment one looks at it. Clearly, the only motivation to entertain this notion is to try to salvage some rather artificial and counterintuitive form of physicalism. And even if such an attempt were to succeed, the world we actually experience would *still* be conditioned by mind, insofar as it would be an outcome of conscious perception. For the purposes of this paper, therefore, the result would be indistinguishable from a truly mental world.

Already in 2005, Johns Hopkins physicist and astronomer Prof. Richard Conn Henry had seen enough. In an essay he penned for *Nature*, he claimed, "The universe is entirely mental. ... There have been serious [theoretical] attempts to preserve a material world—but they produce no new physics, and serve only to preserve an illusion" (Henry 2005: 29). The illusion he was referring to was, of course, that of a world outside mind.

Naturally, Conn Henry's position is controversial and debate around it continues to unfold. Nonetheless, the experiments do show that the idea of a mental world must be taken seriously, if nothing else for the sheer power of the empirical evidence now accumulated. Moreover, philosophers have recently proposed coherent ontologies that can, at least in principle, make sense of reality without the need to postulate anything distinct from mind itself (Kastrup 2017e[3], Nagasawa & Wager 2016, Shani 2015). These ontologies provide coherent frameworks in which the experimental results can be placed and interpreted.

Finally, notice that, although the argument in this section has been based on quantum mechanical experiments carried out on microscopic particles under laboratory conditions, we know that the implications of quantum theory apply to our macroscopic world of tables and chairs as well. Indeed, quantum effects have been experimentally demonstrated for macroscopic objects at room temperature (Lee et al. 2011, Klimov et al. 2015). As such, the failure of non-contextuality indicates that the seemingly objective world we live in is a result of mental process at work

---

3 This reference can be found in Chapter 6 of the present volume.

and, as such, akin to a transpersonal dream: the tables, chairs, stars and galaxies we perceive within it do not have an existence independent of our minds.[4]

15.4 The continuity of mind and world

In a famous paper titled "The Unreasonable Effectiveness of Mathematics in the Natural Sciences," physicist Eugene Wigner (1960) discussed "the miracle of the appropriateness of the language of mathematics for the formulation of the laws of physics." Indeed, abstract methods and results developed purely in thought have, again and again, succeeded in precisely describing concrete phenomena. That axiomatic intuitions turn out to correctly predict and model the structure and dynamics of the world at large is difficult to make sense of under physicalism, this probably being the reason why Wigner used the word 'miracle' twelve times in his paper. After all, lest we incur the fallacy of circular reasoning, under physicalism we cannot logically argue for the validity of logic beyond our own minds, so the world could very well be absurd (Albert 1985). That it is not is Wigner's "miracle."

If the world is mental, however, the correspondence between the intuitive foundations of rational thought and the way the world works is perfectly natural. That we take the basic tenets of logic and mathematics to be self-evident truths betrays their *archetypal nature* in the Jungian sense: they reflect deeply

---

4 As explained in Chapter 6, all *physical* entities, such as tables and chairs, arise within the Markov Blanket of an alter through the inter-action—the interference pattern—between the wave functions corre-sponding to the external state of mind-at-large and the internal state of the alter, respectively. As such, all physical entities depend on the internal state of the alter; that is, the alter's individual mind. The exter-nal state of mind-at-large is *not* a physical state, but a superposition of *thoughts*.

ingrained mental templates according to which thought unfolds (Jung 1991). As a matter of fact, psychologist Marie-Louise von Franz went as far as to argue that the natural numbers themselves are archetypal (1974). Then—and here is the key point—*the fact that these archetypes extend into the world clearly indicates that the world itself is mental and continuous with our minds.* If there is no intrinsic separation between our minds and the objects of perception, naturally these objects should comport themselves in a way consistent with mental archetypes. Perceptual objects should be an expression of archetypal patterns in just the same way that thoughts are, so the world should be consistent—as it is—with our logic and mathematics. The apparent eeriness of Wigner's "miracle" melts away.

To visualize all this consider the following analogy: if mind is like a guitar string, then particular conscious experiences are like particular notes or patterns of vibration of the string. In this case, the mental archetypes discussed above are analogous to the elasticity, mass and length of the string, which determine its normal modes of vibration. Some of the archetypically-defined normal modes of mind thus correspond to the laws of nature, which we discern as regularities on the screen of perception: they reflect some of the 'notes' in which mind naturally 'plays' in the world at large.

Wigner's "miracle" is not only explainable by, but also constitutes further evidence for, the mental world hypothesis. As such, it is high time we considered the implications of this hypothesis for how best to live our lives.

15.5 The implications of a mental world

Strong empirical evidence pointing to the conclusion that the world we experience is a result of transpersonal mental processes at work has now been reviewed. There is no fundamental separation between mind 'in here' and world 'out there,' which explains why the archetypes of rational thought describe nature so well. Yet, the latter point is not the sole implication of a mental world: if our minds are continuous with the environment we inhabit, nothing prevents the world from *intrinsically* evoking mental contents beyond perception, such as insights and emotions.

Indeed, according to analytical psychology, our nightly dreams carry intrinsic semantic value because they are manifestations of deeply ingrained psychological archetypes seeking to express themselves (Jung 1991). By interpreting the archetypal messages our dreams present to us in symbolic form we can, therefore, achieve meaningful insights that escape the reach of ordinary waking introspection (Ackroyd 1993, von Franz & Boa 1994, Jung 2002). Now, if the world is akin to a collective dream also produced by mental archetypes, as discussed in the previous section, *then the same rationale should apply to our waking lives.* The meanings we think to discern in the world may not, after all, be mere personal projections, but actual properties *of the world.* All empirical facts may be archetypal symbols: extrinsic appearances of immanent mental dynamics. The entire cosmic narrative may be hinting at something prior to, and beyond, itself.

In a mental world, the images we perceive on the screen of perception aren't essentially different from our own imagination, except in that the former are shared across observers. This collective 'world dream' symbolically points to underlying *trans*personal mental dynamics, just as regular dreams symbolically point to underlying *personal* mental dynamics. As such, the world is amenable to hermeneutics: it means something; it points to something beyond its face-value appearances; it evokes something *a priori*; it is not its own meaning.

15.6 What the world's traditions have to say

Curiously, despite empirical evidence for the mental world

hypothesis having become available only in relatively recent times, philosophical and spiritual traditions have been hinting at the intrinsic semantic value of the world for millennia. For instance, based on his in-depth study of ancient Islamic mysticism, Henry Corbin suggested that the purpose of life is to interpret the world as a metaphor of transcendent meaning. He wrote:

> To come into this world ... means ... to pass into the plane of existence which in relation to [Paradise] is merely a metaphoric existence. ... Thus coming into this world has meaning only with a view to *leading that which is metaphoric back to true being*. (As quoted in Cheetham 2012: 59, emphasis added.)

That the world isn't literal but metaphorical implies that it isn't the end of a chain of cognitive associations. Instead, its very purpose is to evoke, to point to cognition beyond its face-value appearances.

Analogously, in a clear suggestion that the things and phenomena of the world are symbols of transpersonal mental patterns, Hong Zicheng wrote in the sixteenth century:

> The chirping of birds and twittering of insects are all *murmurings of the mind*. The brilliance of flowers and colors of grasses are none other than the patterns of the Dao. (2006: 105, emphasis added.)

Still along similar lines, the Hermetic tradition suggests that the world is a mental creation in a transpersonal mind:

> That Light, He said, am I, thy God, *Mind ... Mind is Father-God*. ... He [God] *thinketh* all things manifest ... [and] manifests through all things and in all. (Mead 2010: 3, 23, emphasis

added.)

It then proceeds to suggest that the world is the symbolic image of these immanent, transpersonal mental processes:

Holy art Thou, O God ... *of whom All-nature hath been made an image.* (Mead 2010: 11, emphasis added.)

In the West, the inception of these notions goes, of course, all the way back to Plato and his 'Theory of Ideas,' according to which the ontological ground of reality is archetypal thoughts in a transpersonal mind (Ross 1951). The visible world around us is supposedly modeled after the patterns of these archetypal thoughts, which it thus symbolically points to.

Echoing all this, Nisargadatta Maharaj, a twentieth-century exponent of the Advaita Vedanta tradition in India, said:

When you see the world you see God. There is no seeing God apart from the world. Beyond the world to see God is to be God. (1973: 58)

Thus, our only access to God is through the images on the screen of perception that we call the world. These images are the extrinsic appearance of God's conscious inner life. Beyond them, the only way to know God is to gain *direct* access to God's inner life—that is, to *be* God.

I will mention just one more example, since an exhaustive review of how these ideas are represented in the world's traditions is beyond the scope of this brief essay. Christian mystic and scientist Emanuel Swedenborg wrote extensively of the "correspondences" between the natural and spiritual worlds (2007: 63). These correspondences imply that the things and phenomena of the natural world are symbolic images of deeper, transcendent truths. The "correspondences" were Swedenborg's attempt to formulate a hermeneutic of the world.[5]

15.7 Discussion

Physicalism has served important practical purposes over the past couple of centuries. It has provided scientists and engineers with an effective—if simplistic and ultimately wrong—picture of the world, conducive to the development of technology. By thinking of objects and natural phenomena as having standalone reality independent of their own minds, practitioners could achieve the degree of detachment and objectivity necessary for describing the world without bias. The predictive models of nature's behavior that resulted from this effort now lie at the foundation of our technological civilization.

But whilst valuable in a utilitarian sense, this focus on nature's *behavior*—as opposed to nature's *meaning*—is extraordinarily limiting to the human spirit. We are meaning-seeking animals (Frankl 1991, Tillich 1952). A long and productive life enabled

---

5 Here it would have been interesting to mention the vast literature of medieval scholasticism in Europe that resonates directly with the ideas presented in this book. For instance, in his analysis of the thought of medieval scholars, Owen Barfield, based mostly on the writings of Thomas Aquinas, says that, to them, "the world is the thought of God" (2011: 95). Allowing for some leeway regarding the use of the word 'God,' this is precisely a conclusion of Chapter 5. Strikingly, Barfield then adds: "But ... the phenomenon itself [that is, what we experience through sense perception] only achieves its full reality ... in being named or thought by man; for thinking in act *is* the thing thought, in act; just as the senses in act, are the things sensed, in act" (Ibid., original emphasis). This is in precise accordance with the conclusions of Chapter 6 regarding the relational character of the physical world. Barfield removes any possible ambiguity about this by quoting John Scotus Erigena: "The knowledge of things that are, *is* the things" (Ibid.: 99, original emphasis).

by continuous advances in technology is ultimately vacuous and sterile if devoid of meaning. And the same worldview that facilitates the advancement of technology precludes us from finding and appreciating the meaning of life in the world. This, in essence, is perhaps the greatest dilemma of the contemporary zeitgeist.

In such a context, the alternative notion that the world points to something beyond its face-value appearances offers enriching new perspectives. After all, the world we inhabit now carries intrinsic semantic value; a message. Like the Voynich manuscript (Reddy & Knight 2011), it is akin to a book written in a yet-undeciphered language, clamoring for a suitable hermeneutic. Ortiz-Osés' project (2008) turns out to rest on solid metaphysical foundations after all. Each of us, as individuals, can now give ourselves permission to dedicate our lives to *finding meaning in the world*, reassured by the knowledge that this meaning *is really there* even if we can't immediately apprehend it. And whereas the world's meaning won't disappear if we refuse to look for it, the point is that the *option* to look is given legitimacy.

Because of its preoccupation with measurement and predictive modeling, contemporary culture is forgetting to read the letter for the sake of describing the envelope. The physical universe we can measure is merely the carrier of something implied. Exaggerated focus on the predictive models of science, crucial as they are for the development of technology, may distract us from fulfilling what may be our natural and innate telos. In the words of Ortega y Gasset, "Scientific truth is an exact truth, but incomplete and penultimate, that is forcedly integrated in another kind of truth, ultimate and complete yet inexact" (as quoted in Ortiz-Osés 2008: 30).

Looking upon the world interpretatively, as a scholar looks upon an ancient text while trying to decipher its meaning, is not only metaphysically and teleologically sound, it can also make life more wholesome. Psychotherapist Thomas Moore offers us an example: by looking upon our family members as characters and our family stories as episodes of a great saga, meant to subtly evoke something above and beyond its pedestrian literal appearances, we open ourselves up to the deeper archetypal sense they express (2012: 32). By extrapolating this powerful idea further, we can look upon our entire life as a small but crucial element of an unfathomable, symbolic cosmic drama. The experiences we go through are no longer literal and pedestrian, but carry deeper, hidden significance. Indeed, in a mental world it is as unreasonable to interpret life literally as it is to interpret dreams literally. Whoever thinks that a dream is exactly what it appears to be at face value? Most people's instinct upon having an intense dream is to immediately ask themselves: *What does it mean?* Looking upon life in the same way—and asking oneself the same question—can bestow on it a much more spacious, open and wholesome outlook.

With its focus on closed, literal explanations, the physicalist ontology that informs the contemporary zeitgeist decrees that the world has no intrinsic meaning. Instead of an open book waiting to be deciphered and grasped, the world becomes just pixels to be measured; an endless string of quantifiable parameters carrying no message. Instead of the starting point of an open, epic journey along endless cognitive associations, wherein the meanings evoked constitute and ultimately reveal the uncanny reflection of the observer in the observed, the world becomes the end point of a botched quest that never even gets started. By doing this, the physicalist ontology gives us permission to procrastinate in semantic nihilism and an engineered sense of closure. It stops us from pursuing what the Islamic mystics studied by Corbin thought to be the purpose of life. For the ultimate meaning of it all may not be discernible in any particular end point or conclusion, but only in the cognitive gestalt entailed by a *circumambulation*—to use a handy Jungian term—of associative threads. It may be discernible only in a "galaxy" of semantic fields that "are intimately connected, and their significations influence one another, so that the most important sense is found diffuse in its whole" (Ortega y Gasset, as quoted in Ortiz-Osés 2008: 28).

Historically speaking, the denial of the intrinsic symbolic meaning of the world is a recent aberration (Tarnas 2010). The antidote for this aberration is an extension of the application of hermeneutics beyond all discernible boundaries. What we need is a hermeneutic of the entire cosmos; a *Hermeneutic of Everything*.

Closing commentary

Father, O father! What do we here
In this land of unbelief and fear?
The Land of Dreams is better far
Above the light of the morning star.
William Blake: *The Land of Dreams.*

Matter as the outer appearance of inner experience

The entire philosophical system erected in this volume has at its foundation an observation as simple as it is far-reaching: *matter is the outer appearance of inner experience.* This, *and only this,* is what matter is. Nature generously teaches us this lesson every time we look at a living organism's brain: the neural activity we discern is part of what the organism's inner life *looks like* when registered from a second-person perspective; that is, from across a dissociative boundary. The *matter* constituting those neurons is the extrinsic appearance of feeling, emotion, thought, imagination, etc. And since this is what matter *is*, the inanimate universe—also made of matter—must itself be the extrinsic appearance of universal inner life. After all, why would matter be one thing under one set of circumstances—namely, when constituting a living brain—and then something else under another set of circumstances—namely, when constituting the inanimate universe of rocks, clouds and stars? If matter is the outer image of inner experience in certain cases, then—unless we have good reasons to think otherwise, which we don't—it must be the outer image of inner experience in *all* cases. This indicates that the inanimate universe as a whole must be, in a certain sense, akin to a brain. And indeed, as mentioned in Chapter 5, the network topology of the universe at its largest scales does resemble that of a brain (Krioukov et al. 2012); so much so that astrophysicist Franco Vazza and neuroscientist

Alberto Feletti considered the similarity "truly remarkable" and "striking":

> It is truly a remarkable fact that the cosmic web is more similar to the human brain than it is to the interior of a galaxy; or that the neuronal network is more similar to the cosmic web than it is to the interior of a neuronal body. Despite extraordinary differences in substrate, physical mechanisms, and size, the human neuronal network and the cosmic web of galaxies, when considered with the tools of information theory, are strikingly similar. (2017)

Allow me to reiterate my point, in the hope that repetition helps reveal its full force: 'matter' is merely the name we give to the extrinsic appearance of conscious experience, as perceived from across a dissociative boundary. *There is nothing more to it.* This painfully simple insight, repeatedly intimated in nature, is all one needs to come to an understanding of reality that answers all fundamental questions and avoids all fundamental problems, such as the 'hard problem of consciousness' and the 'subject combination problem.' Such an understanding—developed in detail in this book—also makes sense of the experimental results in quantum mechanics that contradict the notion that the world is outside and independent of consciousness. How could an understanding so powerful and so straightforwardly derived from ordinary observation have eluded our culture for so long? After all, it is not like nature has been hiding the clues from us. That we, by and large, have nonetheless failed to see or properly interpret these clues betrays the myriad internalized cultural assumptions and beliefs that bedevil our view of reality.

The ultimate subject as sole ontological primitive

Although the foundational insight discussed above is rather simple, its key implication is perhaps the most difficult nuance to truly grasp in this entire book: consciousness itself—*not an object*—is nature's sole ontological primitive. In other words, literally everything can be reduced to the ultimate *subject* of experience, which isn't anything you can point at. You cannot look at it, for it is that which does the looking. You cannot explain it, for it is that which does the explaining. It is, in fact, *you*; the real 'you' experiencing life through layers of dissociated mentation.

Whereas our cultural indoctrination may render this idea rather elusive, from a strictly analytic or theoretical perspective the choice of consciousness as ontological primitive poses no more problems than alternatives. Allow me to elaborate.

For the idealist ontology proposed in this book to work, we need to imagine that universal consciousness, as ultimate subject, is capable of, and disposed to, *self-excitation*. This disposition to self-excitation is inherent to universal consciousness itself and entails certain *natural modes of excitation*, much like a drumhead vibrates according to certain harmonics and not others. It is self-excitation that allows the structure and complexity of manifest nature to arise from the undifferentiated ground of universal consciousness. Indeed, a simple instrument called a "Chladni Plate" can provide vivid illustrations of how surprisingly complex structure can arise from an undifferentiated substrate through vibration (University of California at Los Angeles, n.d.).

Natural modes of excitation of universal consciousness correspond to the physical laws of nature and to the axiomatic intuitions behind mathematics and logic. As a matter of fact, by reducing both physical regularities (the laws of nature) and psychological regularities (the axiomatic intuitions behind mathematics and logic) to the harmonics of universal consciousness, the ontology articulated in this book explains what Eugene Wigner described as "the miracle of the appropriateness of the language of mathematics for the formulation of the laws of physics" (1960): the truths of human intuition apply to the physical world at large because human intuition and the physical world are, at the most fundamental level, continuous with one another. Physics, mathematics and logic are all archetypal expressions of the ultimate subject in the form of its natural modes of self-excitation.

Moreover, the dynamics of the self-excitations of universal consciousness must ground what we know as 'cognitive associations': the experience of a thought leading to the experience of a memory, which in turn leads to the experience of an emotion, etc. Cognitive associations correspond to the unfolding of excitatory patterns: certain patterns of self-excitation of consciousness lead to—or morph into—other patterns of self-excitation of consciousness, according to some intrinsic affinity. The Chladni Plate can, once again, help visualize this when it depicts transitions between different harmonics in a vibrating surface: each transition is analogous to a cognitive association.

In summary:

(a) Universal consciousness must entail the disposition to self-excitation according to certain natural modes;

(b) The dynamics of transition between its natural modes of self-excitation must ground what we call cognitive associations.

Condition (a) allows us to explain particular qualities of experience as particular patterns of excitation of consciousness. Condition (b) allows us to posit dissociation—the localized *suspension* of the excitatory transitions corresponding to certain cognitive associations—to be a *primary causal phenomenon*, so to explain the emergence of life (see Chapter 5) and of the physical world as seeming object (see Chapter 6). In other words, dissociation is a local change in the configuration of universal consciousness, triggered by primary dispositions of the latter, such that a pattern of self-excitation A in a certain segment of universal consciousness no longer leads to—or morphs into—a pattern of self-excitation B in the same segment of universal consciousness, despite what would otherwise be an intrinsic affinity between A and B.

Now, from an analytic or theoretical viewpoint, does any of this pose any more difficulty than any other choice of ontological primitive? No, it doesn't. Let us take Quantum Field Theory, for instance. According to it, all physical particles—which allegedly constitute the entire universe—consist of particular patterns of self-excitation of a hypothetical, irreducible, underlying quantum field. This quantum field must thus have the fundamental property of being self-excitable according to a variety of modes, which is equivalent to what I've posited for universal consciousness. Moreover, just like universal consciousness, the underlying quantum field cannot be concretely visualized in the absence of its excitations, for in that state—the so-called 'vacuum state'—it is pure potentiality. And finally, again just like universal consciousness, the quantum field is supposedly a fundamental entity that cannot itself be explained in terms of anything else; it simply is. Nothing of what I've posited in this book regarding the properties of universal consciousness as an ontological primitive should sound at all unfamiliar to a quantum field theorist.

And neither should it sound unfamiliar to an M-theorist or a superstring theorist. According to M-theory, our universe consists of patterns of self-excitation of a hypothetical, irreducible, hyper-dimensional membrane—called a 'brane'—vibrating in eleven dimensions. Likewise, superstring theory posits that each physical particle consists of a specific pattern of vibration of an imagined, irreducible entity called a 'superstring.' Naturally, since all measurable or perceivable physical entities supposedly consist in *vibrations* of superstrings or the brane, in the absence of these vibrations neither the brane nor the superstrings can be concretely visualized. Finally, as ontological primitives, superstrings and brane cannot be explained in terms of anything else; they simply are.

By now this should all be sounding very familiar to you. Indeed, every theory of nature must postulate at least one ontological primitive possessing certain dynamical properties. The property of being self-excitable according to a variety of natural modes of vibration is a very popular one in physics, for good theoretical reasons. Hence, my positing a self-excitable universal consciousness as ontological primitive, whose dynamics of excitation ground what we call cognitive associations, poses no extra difficulty in this regard. The only difference is that, instead of postulating an abstract and purely theoretical object as ontological primitive, I chose the primary and sole undeniable datum of existence.

Alternative formulations of dissociation-based idealism

The discussion above provides the framework for the move that allows idealism to be reconciled with a seemingly external, autonomous world: the notion that *dissociation*—a localized blockage in the excitatory dynamics of universal consciousness—is a primary causal phenomenon inherent to the possible behaviors of the ontological primitive. In other words, dissociation is thought to explain life and the world, as opposed to being explained by them. This is a key take-home message of this book.

Indeed, the formulation of idealism presented in Part II is, in principle, not the only one that could be woven around this ground-level notion of dissociation. One could, for instance, conceive of another formulation based on the observation that regular dream images are directly generated by our dreaming psyche—through self-excitation—*already in the very form that they are experienced*. In other words, dream images aren't coded phenomenal representations of some other phenomenal dynamics; they aren't extrinsic appearances of qualitatively different intrinsic views. They are a self-contained movie directly and autonomously generated by our dreaming psyche. So an alternative formulation of idealism could be this: instead of thinking of the inanimate universe as the extrinsic appearance of the conscious inner life of mind-at-large—as posited in Chapters 5 and 6—the corresponding images could be generated at a collective level of universal consciousness, prior to and underlying dissociation, *already in the form we experience them.*

To visualize this, imagine that alters of universal consciousness—that is, living organisms like you and me—are analogous to the seemingly separate branches of a shrub, which ultimately come together at the hidden rhizome. This way, dissociation is the process that creates branches by seemingly separating segments of the shrub. But this process operates somewhat superficially, in that it doesn't affect the unitary rhizome. In this analogy, the inanimate universe we all seem to co-inhabit is a collective dream generated by the rhizome and then broadcast—after some perspectival filtering and adaptation—to all branches already in the form it is experienced. The Jungian notion of a 'collective unconscious' capable of producing archetypal dreams fits nicely with the hypothesis I am trying to describe here: according to it, the world is the waking dream generated by the 'collective unconscious' and then broadcast to each of our individual psyches.

This isn't as far-fetched as it may sound at first. Indeed, as discussed in Chapter 9, there is an empirically known form of dissociation according to which subjects lose the sense of ownership of their own mental contents (Klein 2015). In this context, it is not unreasonable to imagine that empirical reality is a collective stream of imagination that we lose our sense of ownership of, thereby mistakenly concluding that it corresponds to some external world.

Nonetheless, such a seemingly elegant formulation of

idealism fails because, whereas it can parsimoniously explain the inanimate universe, it cannot explain the presence of other conscious organisms in it. If the collective dream we call 'the world' were broadcast from the rhizome to the individual branches, why or how would one branch experience the presence of other branches—that is, other people and living organisms—in its dream? After all, two TV receivers tuned to the same channel can display the same movie, *but not images of each other within that movie.*

You see, for the same reason that this alternative formulation parsimoniously does away with the conscious inner life of mind-at-large, it must also do away with the conscious inner life of other living organisms. This effectively reduces it to solipsism and renders redundant the very need to explain a shared world to begin with.

I have taken the trouble to explore this defunct formulation here for the following reason: in my experience, many intelligent people who hear the idealist adage 'reality is akin to a collective dream' try to construe its meaning along the general outline of this formulation. My hope is that the brief discussion above anticipates and preempts this futile labor. As seen in Chapter 8, idealism is different from solipsism and can be articulated in a distinctly coherent manner.

The linguistic problem of apparent contradictions
The key difference between the defunct formulation of idealism discussed above and that elaborated upon in Part II is this: the former entails that the images on our personal screen of perception are themselves irreducible. The latter, on the other hand, posits that personal perceptions can be partially reduced to the phenomenality of mind-at-large—in the sense that personal perceptions are generated by interaction between the thoughts of the alter and those of mind-at-large—which in turn is *qualitatively different from personal perceptions.*

Bishop Berkeley's formulation of idealism is similar to that discussed in the previous section, insofar as it also entails that personal perceptions are irreducible. As explained by Barfield, "Berkeley held that ... the representations *as such* [that is, personal perceptions], are sustained by God in the absence of human beings" (2011: 36, original emphasis). Even some present-day academic philosophers continue to entertain the idea that the contents or qualities of personal perception are themselves irreducible: "In perception, our finite unities of consciousness come to literally overlap with the unity of consciousness that is reality" (Yetter-Chappell forthcoming). So the phenomenal properties you experience when you see the world are supposedly *the same phenomenal properties* encompassed by "the unit of consciousness that is reality," with which you overlap during the act of perceiving.

I believe that all formulations of idealism entailing such irreducibility of personal perceptions fail, either because of the difficulties discussed in the previous section or at least because of the consequences of trying to circumvent these difficulties. However, the point I want to make now is different: by maintaining that personal perceptions can be partially reduced to something outside the personal self—that is, outside alters of universal consciousness—I am positing something at least analogous to what Barfield called the "unrepresented" (2011) and Kant the "noumenal." Indeed, I am maintaining that *there is a shared reality beyond the alters*—namely, the thoughts of mind-at-large in a state of superposition (see Chapter 6)—underlying our personal perceptions of the world. This shared reality would still exist even if we and all other living beings ceased to be. So the noumenal does exist, my key point being simply that its underlying nature is phenomenal. After all, the noumenal itself consists of experiences, even though these experiences are qualitatively different from personal perceptions. In summary, according to the ontology defended in this book, *there are noumena but they are phenomena.*

There is no denying that, at first sight, such a statement appears to be a contradiction. But appearances are misleading, as readers will hopefully see at this stage of my argument. Indeed, Kant and Barfield expressly did *not* specify the ontological character of the noumenal and the unrepresented, respectively, so neither is necessarily dichotomous with the phenomenal. The problem is that, without the semantic context provided by the preceding chapters, it would be tricky to defend the coherence of my apparently contradictory statement. The phenomena that make up the noumena are not phenomena of alters, but of mind-at-large. How to make this clear and compelling without the extensive elaboration on universal consciousness, alters and mind-at-large in Part II?

Unfortunately for the idealist, language difficulties do not end here. For example, in public talks and presentations I often claim that, unlike what physicalism and dualism imply, *the world is exactly what it seems to be: it consists of qualities of experience.* Without semantic context, it would be very easy—even forgivable—to construe this statement to mean a Berkeleyan form of idealism, whereby the *shared* world beyond your personal self comprises the qualities of your *personal* perceptions, as discussed above. Yet, this would contradict the ontology articulated in Part II. What I actually mean by the statement is one of two things: depending on the semantic context, I may mean that our *shared* world—that is, the thoughts of mind-at-large, which are *not* physical—consists of qualities of experience, *but not the qualities that you experience personally through perception.* Alternatively, I may mean that the *physical* world—that is, the world of tables and chairs around you, which is not shared but *private*, as discussed in Chapter 6—is exactly what it seems to be, in the sense that it consists of the qualities of *your personal* perception, brought into being in the Markov Blanket of *your* alter, as a result of *your* interaction with mind-at-large. Do you see the tricky but unavoidable nuance associated with each case?

I shall mention one more example, which is perhaps the most notorious language difficulty for the idealist: the usage of the pronouns 'I' and 'you.' As we have seen, according to the ontology defended in this book each person is an alter of universal consciousness. In this context, 'I' and 'you' refer to *different alters.* However, I have also argued that the essential subjectivity of each alter is that of universal consciousness itself. In this alternative context, 'you' is universal consciousness and so is 'I.' This may be culturally unusual but, according to my argument, it is more accurate. So if you asked me, for instance, 'Bernardo, will I cease to exist when my physical body dies?' both affirmative and negative answers would be equally valid, depending on what you mean by 'I': if 'I'—that is, you—is universal consciousness, then the answer would be 'No, death happens within that which you are.' But if 'I' is your alter, then the answer would be 'Yes, what we call death is the extrinsic appearance of the dissolution of the dissociative boundary of your alter.'

In practice, the idealist will be often forced to implicitly assume a particular meaning for the pronouns 'you' and 'I' during discussions. I, for one, make this choice based on the semantic context I believe to be most appropriate for the particular exchange at hand. For instance, in non-dualist philosophy circles people often take 'you' and 'I' to refer to one's essential subjectivity, or universal consciousness. In analytic philosophy discussions, on the other hand, 'you' and 'I' are always taken to refer to the personal self, or alter. Therefore, I am certain that I could be correctly quoted as having said both that 'You will not survive the death of your body' *and* 'You will certainly survive the death of your body.' Does this mean that I—that is, this alter called Bernardo Kastrup—am confused and my views are contradictory? Does this place an obligation on me to *define* what I mean by the pronouns 'I' and 'you'—as I have just done in the preceding sentence, to cumbersome effect—*each time* I use them?

There are many more instances of situations wherein an uncharitable audience or interlocutor may conclude that the idealist's defense of his views is plagued by contradictions. With the three examples above, I hope to have illustrated how important it is for the idealist to have an opportunity to establish the semantic context of his argument adequately, before conclusions can be drawn about it. Yet, such an opportunity is a rare privilege in this age of impatient pill-form communications, when one is expected to make one's point merely through brief aphorisms. There is precious little tolerance for thoughtfulness and nuance nowadays, and an eagerness to point one's accusatory finger at the slightest appearance of contraction.

Language now carries so much cultural baggage that the idealist must explicitly and laboriously define his semantic context even when it seems self-evident. It is essential that not only the idealist, but also his audience, be aware of this inherent difficulty. Those sincerely committed to the pursuit of truth will listen charitably to the idealist's arguments, seeking what is *meant* above what is merely *said*. After all, the pursuit of truth is not a game between rival factions; its purpose is not to defeat the opposition, but to understand what is valid and meaningful *in one's own life*. Those who neglect this kind of self-honesty will succeed only in deceiving themselves.

The unavoidable conundrum of spacetime

Another issue associated with—though broader than—language is spacetime. Indeed, space and time are built not only into language—think of tenses, for instance—but also our *very way of thinking*. Perhaps we think in terms of spacetime because we think grammatically, or perhaps spacetime is built into language because language mirrors the way we think; I do not know; probably both. But the answer is irrelevant for my purposes here. The fact is—and this should be self-evident enough—that any statement about what nature is or how it works presupposes a spacetime scaffolding. Without extension in at least one dimension, the various states of nature would overlap and become indistinguishable from one another. Information about nature—as defined by Shannon (1948)—would thus vanish and there would remain literally nothing to be said about it.

My earlier analogy between experiences and vibrations of consciousness presupposes a spacetime scaffolding circumscribing consciousness. After all, vibrations entail some form of movement in space and time (think of a guitar string playing a musical note: it moves up and down as time passes). So it could be argued that the idealist ontology defended in this book, in addition to consciousness itself, assumes a spacetime scaffolding as extra primitive, wherein consciousness can then 'move' so to have or produce particular experiences. But this would contradict my core claim that consciousness is the sole primitive. Similarly, when I discussed the ultimate goal or telos of life in Chapter 15, I implicitly assumed an objective time dimension. After all, goals presuppose an extension of reality from present to future. But this, too, would appear to contradict the core claim of this book.

So let me be clear: my position is that both space and time are *qualities of experience*. Time exists only insofar as what we call 'past' is an experiential quality characteristic of memory and 'future' an experiential quality characteristic of imagined possibilities or expectations. Space, in turn, exists only insofar as it is the experiential quality of a certain *relationship* between perceived objects. So spacetime is not an ontological primitive; it isn't some kind of scaffolding independent of consciousness. It is only an amalgamation of qualities—amenable to mathematical modeling—that themselves exist only *in* consciousness. I thus stand by my core claim that consciousness is the sole ontological primitive.

But now I must somehow reconcile this core claim with telos

and my earlier analogy between experiences and vibrations of consciousness. To begin with, the analogy must be regarded solely as such: as an *analogy*. This way, experiences are *like* vibrations of consciousness. The intent of the analogy was to help you visualize how various experiences can be distinct from each other without requiring that there be anything to them but consciousness itself. As a matter of fact, I defined experiences as *excitations* — as opposed to outright vibrations — of consciousness in the hope that the term 'excitation' wouldn't commit me as much to dimensional extension.

You see, the problem is that if I — or anyone else, for that matter — want to *talk* about the nature of reality, I must presuppose a spacetime scaffolding at least metaphorically. This is a *concession* to the limitations of thought and language. Nonetheless, acknowledging that this dimensional scaffolding is simply a kind of illusion inherently imposed by the structure of our cognition doesn't change the practical problem at hand: whatever reality precedes spacetime ontologically is unreachable by the human intellect. At best, we can cognize *projections* of this otherwise inaccessible reality onto the cognitive scaffolding of spacetime.

Here is an analogy to help you see what I mean: we cannot read a letter written in a piece of paper that has been folded multiple times over into a small, nearly dimensionless crumple. The characters overlap and the information they contain becomes indiscernible. Only by unfolding the crumple — thereby *extending* the paper — can we make sense of the message it contains. Reality prior to spacetime is, in a sense, like the paper crumple: we need to unfold it along the dimensions of space and time to render it amenable to intellectual apprehension.

Does this mean that our spacetime-bound thought processes can never arrive at valid and meaningful conclusions? That notions such as telos, purpose, movement and vibration are delusional and carry no truth? No. All it means is that the notions we construct and the conclusions we reach within the framework of spacetime cannot be *ultimately* true. After all, *ex hypothesi*, spacetime is merely a cognitive construction. However, they can still be *penultimately* true in the sense that they can accurately *correspond* to something prior to dimensionality. Valid spacetime-bound conclusions are thus *projected images* of ultimate truths, adapted to the limitations of human cognition by dimensional extension (that is, by the unfolding of the otherwise 'dimensionless' crumple).

This way, to say that experiences are vibrations of consciousness admittedly cannot be ultimately true, for consciousness — as sole ontological primitive — does not occupy a spacetime scaffolding prior to itself. But it can still be true in the penultimate sense that vibrations *correspond* to something true — though ineffable — about consciousness prior to dimensional extension; that vibrations are an accurate *projected image* of what ultimately happens in consciousness when it experiences. That we cannot directly think or say something coherent about an ultimate truth does not invalidate our penultimate images and concepts. They can still tell us something *in*directly true about what reality is and how it works.

As a matter of fact, space itself can be coherently regarded as the aspect of human experience that corresponds most closely to universal consciousness, different segments of the latter corresponding to different regions of space. Indeed, that two living organisms — the extrinsic appearances of alters — never occupy the same volume of space reflects the notion that different alters are located in different segments of universal consciousness, as discussed in Part II. Moreover, that we think of empty space as a void, a nothing, reflects the notion that unexcited universal consciousness cannot, by definition, be experienced. Even the idea that unexcited universal consciousness must have intrinsic properties — otherwise there would be nothing to eventually *get* excited — finds a correspondence in how we think of space at least since the early twentieth century: empty space, too, is a void with intrinsic properties. The correspondences here are clear: there is a strong sense in which, as far as human cognition is concerned, empty space *is* universal consciousness, the contents of space being excitations of universal consciousness. Moreover, since space is simply a facet of spacetime, I suggest that it is closer to the truth to think of spacetime *as* universal consciousness than as a scaffolding *occupied by* universal consciousness.

With regard to the temporal dimension, the existence of telos may also be a penultimate truth: an *image* in time of something timeless. But the image is still valid as such. My claim is thus that there is something timeless in nature that, from within the framework of spacetime, appears to us as telos or purpose. If so, then stating that life and the universe have a telos is as true and meaningful an assertion as can be made from within the confines of human cognition. Saying otherwise would, in turn, be outright false.

If one is to make and talk about philosophy, I believe it is unavoidable to frame one's thoughts and discourse in terms of spacetime extension. For the true idealist, this is admittedly a concession, for spacetime is not in the idealist's reduction base. The assertions made should thus *not* be regarded as ultimate. But they are still true and meaningful as far as they go. The alternative is to remain silent while ontological nonsense continues to sweep our culture.

Cultural indoctrination precludes knowledge by acquaintance

Although the limitations of logical thought and language are serious challenges for the idealist, the greatest difficulty even educated people have with idealism is more related to intuitive feeling than reasoned argument. You see, it can be argued—as discussed in Chapter 3—that all non-idealist ontologies run into insoluble problems, such as the 'hard problem of consciousness' and the 'subject combination problem.' All non-idealist ontologies fail to explain some key and undeniable aspect of nature. And all non-idealist ontologies, in addition to these limitations, *also* fail the parsimony test by postulating all kinds of thought abstractions — conceptual entities whose actual existence is fundamentally impossible to verify empirically in any direct manner — as ontological primitives. When these observations are truly taken to heart, one wonders why there is even debate about the superior adequacy of idealism. The psychological motivations behind physicalism — discussed in Chapter 14 — do help explain the historical inception and subsequent momentum gathered by our mainstream ontology, but they aren't sufficient to explain why you, dear reader, still look around and — despite what your intellect may be telling you — *still* see a world outside mind. Try it: lift your head from this page and look around right now. Do you sincerely see only mind at work?

Unfair as this may be to some of you, it is probably safe to say that most people, convinced as they may be of my argument at the level of rational thought, still can't *feel* the world to be a mental unfolding. In all honesty, most days I can't either. But when I do, the answer to the riddle of ontology becomes self-evident. Even the theoretical argument developed in this book becomes redundant, for the felt apprehension of the mental universe — the knowing of it by acquaintance — is vastly more compelling than conceptual discourse.

The problem — as discussed in Chapter 7 — is that educated humans can hardly see the world for what it is, but instead see it as culture has taught them to. We have been culturally indoctrinated for centuries to see the world as something outside and independent of mind, so that's what we see when we look around.

As I have pointed out before (Kastrup 2015: 142-146), to break this formidable spell we need more than a solid intellectual argument. *But a solid intellectual argument is still an essential enabler.* Allow me to elaborate.

The role of academia in the cultural dialogue

In our telecommunications age, the evolution of culture is at least as much a top-down process as it is a bottom-up one. In other words, significant aspects of our culture's worldview are defined by intellectual elites—namely academia—and then disseminated throughout society by the mainstream media. Alternative and social media, whose advent initially promised to reduce our cultural dependency on the often-calcified views of academic elites, are now perhaps doing more harm than good: the fact-free, hysterical nonsense that often comes through these channels prevents them from garnering credibility. As such, it is unlikely that the cultural influence of the elites and the mainstream media will be much diminished in the foreseeable future.

And here is where a solid intellectual argument is essential for healthy changes in our culture's understanding of reality. Having explored alternatives, I have come to the tentative conclusion that these changes must start in academia itself, despite its often-calcified stances. Indeed, as Thomas Kuhn lucidly observed (1996), it is a formidable challenge to reform the fundamental assumptions—the "paradigm"—according to which academia operates. Yet, I now consider it still more achievable than the alternative: that of reforming the mechanisms governing the cultural evolution of our society, including its established networks of trust and structures of communication. If academia embraces a saner, more mature ontology, lending it its credibility, this change will quickly cascade down to society at large through established, well-oiled mechanisms. And, as discussed in Chapter 15, such a change is urgently needed to bring not only truth, but also meaning back into our lives.

My hope for this book

In this context, this book represents my attempt to reach out to academia in an admittedly critical but—hopefully—also well-grounded and well-argued manner. I hope this contribution helps to set off or accelerate a sincere and open-minded reevaluation of our options regarding ontology, as well as to reform the way we think about ontology. Ultimately, there is much to be gained for all involved.

In my attempt to construct a gap-free argument, I have had to dabble in multiple fields of academic inquiry in which I am not an authority. After all, an appropriate proposal for a new *world*view must touch on all salient aspects of the *world*. The problem, of course, is that the world doesn't comply with the neat boundary lines that separate different academic disciplines. In it, there is no neuroscience as distinct from psychology, or philosophy as distinct from physics. For this painfully simple reason—which much of academia may have conveniently overlooked for the past century or two—an appropriate case for an ontology must be multi-disciplinary. There is no escaping this, despite the fact that myriad ontology papers in academic journals continue to focus exclusively on conceptual minutia and abstract formalisms. If we want to explain the facts of the world, we must remain judiciously acquainted with the latest relevant knowledge of these facts, whatever academic discipline they pertain to. Moreover, the same rationale applies the other way around: *facts are legitimate building blocks for an ontological argument even when they are not yet fully understood from an analytic standpoint.* For instance, the formation of co-conscious alters in dissociative identity disorder doesn't cease to be an empirical fact simply because we may not yet be able to unpack precisely how this dissociation works. As such, dissociation is a valid building block for an ontology even if its inner workings remain, as yet, partly mysterious.

To write this book I have had to venture not only into

philosophy, but also psychology, psychiatry, neuroscience, physics and even biology. In doing so, I am bound to have missed relevant literature or worded my arguments in a way that experts in the respective disciplines might consider inadequate, peculiar or even naïve. Moreover, in my attempt to make my case understandable across disciplines by building it up from first principles, I am also bound to have done insufficient justice to related work by other scholars. I can only hope that experts will consider my case charitably, focusing on its essence, as opposed to its form, ethical gaffes and ancillary details. If certain steps of my overall argument are found to lack sufficient rigor, I hope this won't discourage my peers from still considering the argument *as a whole*. I trust this book has enough substance to be regarded as offering at least an intriguing hypothesis.

And often an intriguing hypothesis is all that is needed to liberate the imagination—even the analytic or theoretical imagination. Viable possibilities and avenues of thought must be first *imagined* if one hopes to break out from ingrained assumptions and patterns of thinking. As such, if nothing else, I hope this book succeeds in opening up new paths of inquiry around the twin notions of (a) spatially-unbound, self-excitable consciousness as sole ontological primitive; and (b) mental dissociation as a primary causal mechanism in nature.

Despite the failure of physicalism, we are not at a dead-end in terms of ontology, but at a crossroads instead. Choosing the new ontological foundation for our relationship with nature and each other—based on reason, parsimony and empirical adequacy—is the challenge and responsibility incumbent upon academia. And by this I mean that an actual *choice* must be made, or at least an unambiguous *preference* stated, as opposed to an endless state of politically correct, collegial 'openness' to all alternatives on the table, which may secure smooth careers but brings our culture precisely nowhere. *We know enough to choose deliberately*, as opposed to allowing the momentum behind historical undercurrents—outside the control of reason and uninformed by the latest evidence—to perpetuate an untenable view of reality. Our collective sanity and vitality depend on this deliberate choice.

Afterword by Edward F. Kelly: Science and spirituality: An emerging vision

After we came out of the church, we stood talking for some time together of Bishop Berkeley's ingenious sophistry to prove the nonexistence of matter, and that every thing in the universe is merely ideal. I observed, that though we are satisfied his doctrine is not true, it is impossible to refute it. I never shall forget the alacrity with which Johnson answered, striking his foot with mighty force against a large stone, till he rebounded from it—'I refute it thus.'
James Boswell: *The Life of Samuel Johnson.*

Kick at the rock, Sam Johnson, break your bones; but cloudy, cloudy is the stuff of stones.
Richard Wilbur: *Epistemology.*

The rise of modern science has brought with it a host of extraordinary intellectual and practical achievements, but a host of serious and worsening problems as well. Many if not all of these problems seem connected somehow with a deep split that has developed in modern times between science and spirituality. This split itself resulted mainly from the recent ascendance of *secular humanism*, a worldview that is anchored in the classical physical science of the late 19th century and profoundly hostile to all things religious, in which it sees only vestiges of our intellectual childhood. This 'physicalist' worldview basically holds that reality consists at bottom of tiny bits of solid self-existent stuff moving in accordance with mathematical laws under the influence of fields of force, and that everything else, including our human minds and consciousness, must emerge somehow from that basic stuff. Our everyday understanding of ourselves as effective conscious agents equipped with free will is delusive, because we are in fact nothing more than extremely complicated biological machines. Consciousness and its contents are generated by (or in some mysterious way identical to, or supervenient upon) neurophysiological processes in the brain. Beliefs about postmortem survival, common to the world's religious traditions, are therefore also delusive: Biological death is necessarily the end, because without a functioning brain there can be no mind and consciousness, period. On a more cosmic scale, we see no sign of final causes or a transcendent order. The overall scheme of nature appears utterly devoid of meaning or purpose.

Views of this sort have permeated the opinion elites of all advanced societies and undoubtedly contribute to the pervasive 'disenchantment' of the modern world with all of its accompanying ills. They have also accumulated enormous cultural momentum and become in effect self-perpetuating by gaining near-total control of key elements of modern society such as our educational institutions and the media. In recent decades our secondary schools, colleges, and universities have all in effect become advocates for the prevailing physicalist worldview, which by now not only dominates mainstream *scientific* disciplines such as biology, neuroscience, cognitive psychology and the social sciences, but also has destructively colonized neighboring academic areas including the humanities generally (perhaps most surprisingly, religious studies), and even theology. It has also encouraged the recent spate of scientistic attacks on traditional religions, especially the Abrahamic religions, which in turn has engendered pushback in the various forms of fundamentalist fanaticism we witness with depressing regularity on the evening news.

Classical physicalism, however, is not merely incomplete, but incorrect at its very foundation. The deterministic clockwork universe postulated by Newton and Laplace was overthrown with the rise of quantum theory a century ago, and 'matter' as classically conceived shown not to exist. Contemporary physicalist brain/mind theory is headed in the same direction. At present we have no understanding whatsoever of how consciousness could be generated by physical events in brains, and recent theoretical work in philosophy of mind has convinced many that we can never achieve one. Meanwhile, large amounts of credible empirical evidence have accumulated for a variety of human mental and psychophysical capacities that resist or defy explanation in conventional physicalist terms. These 'rogue' phenomena include, for example, paranormal or 'psi' abilities of various kinds, extreme forms of psychophysical influence such as stigmata and hypnotic blisters, the most basic experiential properties of our human memory system, multiple and overlapping centers of consciousness associated with single physical organisms, powerful near-death experiences occurring under extreme physiological conditions such as deep general anesthesia and/or cardiac arrest, genius-level creativity, and mystical experiences whether spontaneous, the result of intensive meditative practice, or induced by psychedelics. There is even direct evidence of several substantial kinds for postmortem survival, coupled with increasing recognition that the only credible explanations for this evidence involve either survival itself or psi processes in and among living persons—a dilemma both horns of which are fatal to the physicalist worldview.

Classical physicalism is too impoverished to carry this heavy empirical burden, but what should take its place? Serious attempts to imagine how reality must be constituted, in order that rogue phenomena of the indicated sorts can happen, appear to lead inescapably into metaphysical territory partially overlapping with the world's spiritual traditions—specifically, toward some yet-to-be-fully-characterized form of *evolutionary panentheism*. An idealist worldview of this type rests upon just three core principles: First, that the manifest world arises from and is constituted by a tremendous world-transcending ultimate reality of some conscious sort; second, that we humans are intimately linked with that ultimate reality in the depths of our individual psyches, and can experience it directly in a variety of ways; and third, that the antecedently existing universal consciousness or universal self that is the source of the manifest universe is in some sense slowly waking up to itself as evolution of more complex biological forms enables fuller expression of its inherent capacities.

What is currently emerging, in short, is a middle way between the warring fundamentalisms—religious *and* scientific— that have dominated recent public discourse; specifically, an expanded science-based understanding of nature that can accommodate empirical realities of spiritual sorts while also rejecting rationally untenable 'overbeliefs' of the sorts targeted by the strident contemporary critics of institutional religions. This emerging vision is both scientifically justifiable and spiritually satisfying, combining the best aspects of our scientific and religious heritage in an intellectually responsible effort to reconcile these two greatest forces in human history. It can provide sustenance in particular to persons who view themselves as 'spiritual but not religious,' and to those who remain in a traditional faith but are troubled by inescapable conflicts between elements of religious doctrine and the teachings of science. At the same time, like traditional faiths, it makes room for the possibility of postmortem survival and can therefore provide comfort to persons who are facing the reality of death, whether for themselves or for loved ones such as aging parents, or who have themselves encountered powerful mystical-type experiences through meditation, psychedelics, or a close brush with death.

The vision sketched here provides an antidote to the prevailing postmodern disenchantment of the world and demeaning of human possibilities. It not only more accurately and fully reflects our human condition but engenders hope and encourages ego-surpassing forms of human flourishing. It offers reasons for us to believe that freedom is real, that our human choices matter, and that we have barely scratched the surface of our human potentials. It also addresses the urgent need for a greater sense of worldwide community and interdependence—a sustainable *ethos*—by demonstrating that under the surface we and the world are much more extensively interconnected than previously recognized.

Our individual and collective human fates in these dangerous and difficult times—indeed, the fate of our precious planet and all of its passengers—may ultimately hinge upon wider recognition and more effective utilization of the higher states of being that are potentially available to us but largely ignored or even actively suppressed by our postmodern civilization with its strange combination of self-aggrandizing individualism and fundamentalist tribalisms. Availability of an improved worldview does not guarantee its acceptance, of course, and even widespread acceptance would not guarantee that its potential benefits will be fully realized, or its potential abuses adequately controlled. But a viable pathway to a better world does appear in principle to be opening up. Bernardo Kastrup has contributed at many levels to the development of this emerging vision, and this fine new book gives me real hope that the main barrier to its widespread acceptance—exemplified in Samuel Johnson's response—is on the verge of collapse. A major inflection point in modern intellectual history is close at hand!

Edward F. Kelly, PhD is a Professor in the Department of Psychiatry and Neurobehavioral Sciences at the University of Virginia, and lead author of Irreducible Mind *(2007) and* Beyond Physicalism *(2015).*

Appendix: The idealist view of consciousness after death

This article first appeared in the *Journal of Consciousness Exploration & Research*, ISSN: 2153-8212, Vol. 7, No. 11, pp. 900-909, in December 2016. It was an invited contribution to a special issue of the journal and did not go through peer-review (only editorial review). As such, the article did not meet the criteria for inclusion in the main body of this volume. Yet, the subject matter addressed in it is squarely within the scope of this work, justifying its inclusion as an appendix.

A.1 Abstract

To make educated guesses about what happens to consciousness upon bodily death, one has to have some understanding of the relationship between body and consciousness during life. This relationship, of course, reflects an ontology. In this brief essay, the tenability of both the physicalist and dualist ontologies will be assessed in view of recent experimental results in physics. The alternative ontology of idealism will then be discussed, which not only can be reconciled with the available empirical evidence, but also overcomes the lack of parsimony and limited explanatory power of physicalism and dualism. Idealism elegantly explains the basic facts of reality, such as (a) the fact that brain activity correlates with experience, (b) the fact that we all seem to share the same world, and (c) the fact that we can't change the laws of nature at will. If idealism is correct, the implication is that, instead of disappearing, conscious inner life expands upon bodily death, a prediction that finds circumstantial but significant confirmation in reports of near-death experiences and psychedelic trances, both of which can be construed as glimpses into the early stages of the death process.

A.2 Introduction

Our capacity to be conscious subjects of experience is the root of our sense of being. After all, if we weren't conscious, what could we know of ourselves? How could we even assert our own existence? Being conscious is what it means to be us. In an important sense—even the only important sense—we are first and foremost consciousness itself, the rest of our self-image arising afterwards, as thoughts and images constructed in consciousness.

For this reason, the question of what happens to our consciousness after bodily death has been central to humanity throughout its history. Do we cease to exist or continue on in some form or another? Many people today seek existential solace in body-self dualism, which opens up the possibility of the survival of consciousness after bodily death (Heflick et al. 2015). But is dualism—with the many serious problems it entails, both philosophical and empirical (Robinson 2016)—the only ontology that allows for this survival?

Although consciousness itself is the only directly accessible datum of reality, both dualism and the mainstream ontology of physicalism (Stoljar 2016) posit the existence of something ontologically distinct from consciousness: a physical world outside and independent of experience. In this context, insofar as consciousness is believed to be constituted, generated, hosted or at least modulated by particular arrangements of matter and energy in the physical world, the dissolution of such arrangements—as entailed by bodily death—bears relevance to our survival. This is the root of humanity's preoccupation with death.

However, the existence of a physical world outside and independent of consciousness is a theoretical inference arising from interpretation of sense perceptions, not an empirical fact. After all, our only access to the physical is through the screen of perception, which is itself a phenomenon of and in consciousness.

Renowned Stanford physicist Andrei Linde summarized this as follows:

> Let us remember that our knowledge of the world begins not with matter but with perceptions. ... Later we find out that our perceptions obey some laws, which can be most conveniently formulated if we assume that there is some underlying reality beyond our perceptions. This model of material world obeying laws of physics is so successful that soon we forget about our starting point and say that matter is the only reality, and perceptions are only helpful for its description. This assumption is almost as natural (and maybe as false) as our previous assumption that space is only a mathematical tool for the description of matter. (1998: 12)

The physical world many believe to exist beyond consciousness is an abstract explanatory model. Its motivation is to make sense of three basic observations about reality:

(a) If a physical brain outside experience doesn't somehow generate or at least modulate consciousness, how can there be such tight correlations between observed brain activity and reported inner experience (cf. Koch 2004)?

(b) If the world isn't fundamentally independent and outside of experience, it can only be analogous to a dream in consciousness. But in such a case, how can we all be having the same dream?

(c) Finally, if the world is in consciousness, how can it unfold according to patterns and regularities independent of our volition? After all, human beings cannot change the laws of nature.

Nonetheless, if these questions can be satisfactorily answered without the postulate of a physical world outside consciousness, the need for the latter can be legitimately called into question on grounds of parsimony. Moreover, while physicalism requires the existence of ontological primitives — which Strawson called "ultimates" (2006: 9) — beyond consciousness, it fails to explain consciousness itself in terms of these primitives (cf. Chalmers 2003). So if the three basic observations about reality listed above can be made sense of in terms of consciousness alone, then physicalism can be legitimately called into question on grounds of explanatory power as well. And as it turns out, there is indeed an alternative ontology that explains all three basic observations without requiring anything beyond consciousness itself. This ontology will be summarized in Section A.4 of this brief essay.

In addition, the inferred existence of a physical world outside and independent of consciousness has statistical corollaries that can be tested with suitable experimental designs (Leggett 2003, Bell 1964). As it turns out, empirical tests of these corollaries have been carried out since the early eighties, when Alain Aspect performed his seminal experiments (Aspect, Grangier & Roger 1981). And the results do not corroborate the existence of a universe outside consciousness. These seldom-talked-about but solid empirical facts will be summarized in the next section.

Without a physical world outside consciousness, we are left with consciousness alone as ground of reality. In this case, we must completely revise our intuitions and assumptions regarding death. After all, if consciousness is that within which birth and death unfold as phenomenal processes, then neither birth nor death can bear any relevance to the existential status of consciousness itself. What does death then mean? What can we, at a personal level, expect to experience upon bodily death? These questions will be examined in Section A.5 of this essay.

A.3 The empirical case against a world outside consciousness

A key intuitive implication of a world outside consciousness is that the properties of this world must not depend on observation; that is, an object must have whatever properties it has—weight, size, shape, color, etc.—regardless of whether or how it appears on the screen of perception. This should clearly set the physical world apart from the sphere of consciousness. After all, the properties of a purely imagined object do not exist independently, *but only insofar as they are imagined.*

As mentioned earlier, the postulated independence of the world from observation has certain statistical corollaries (Leggett 2003) that can be directly tested. On this basis, Gröblacher et al. (2007) have shown that the properties of the world, surprisingly enough, *do* depend on observation. To reconcile their results with physicalism or dualism would require a counterintuitive redefinition of what we call *objectivity.* And since contemporary culture has come to associate objectivity with reality itself, the science press felt compelled to report on this study by pronouncing, "Quantum physics says goodbye to reality" (Cartwright 2007). Testing similar statistical corollaries, another experiment (Romero et al. 2010) has confirmed that the world indeed doesn't conform to what one would expect if it were outside and independent of consciousness.

Other statistical corollaries (Bell 1964) have also been experimentally examined. These tests have shown that the properties of physical systems do not seem to even exist prior to being observed (Lapkiewicz et al. 2011, Manning et al. 2015). Commenting on these results, physicist Anton Zeilinger is quoted as saying that "there is no sense in assuming that what we do not measure about a system has [an independent] reality" (Ananthaswamy 2011). Finally, Ma et al. (2013) have again shown that no naively objective view of the world can be true.

Critics have deeply scrutinized the studies cited above to find possible loopholes, implausible as they may be. In an effort to address and close these potential loopholes, Dutch researchers performed an even more tightly controlled test, which again confirmed the earlier results (Hensen et al. 2015). This latter effort was considered the "toughest test yet" (Merali 2015).

Another intuitive implication of the notion of a world outside consciousness is that our choices can only influence the world—through our bodily actions—in the present. They cannot affect the past. As such, the part of our story that corresponds to the past must be unchangeable. Contrast this to the sphere of consciousness wherein we can change the whole of an imagined story at any moment. In consciousness, the *entire* narrative is always acquiescent to choice and amenable to revision.

As it turns out, Kim et al. (2000) have shown that observation not only determines the physical properties observed at present, *but also retroactively changes their history accordingly*. This suggests that the past is created at every instant so as to be consistent with the present, which is reminiscent of the notion that the world is a malleable mental narrative.

Already back in 2005, renowned Johns Hopkins physicist and astronomer Richard Conn Henry penned an essay for *Nature* wherein he claimed that "The universe is entirely mental. ... There have been serious [theoretical] attempts to preserve a material world—but they produce no new physics, and serve only to preserve an illusion" (2005: 29). The illusion he was referring to was, of course, that of a world outside consciousness.

Thus from a rigorous empirical perspective, the tenability of the notion of a world outside and independent of consciousness is at least questionable. The key reason for resisting an outright abandonment of this notion is the supposed lack of plausible alternatives. What other ontology could make sense of the three basic observations about reality discussed in Section A.2? In the next section, I will attempt to answer this question.

A.4 A simple idealist ontology

The ontology of idealism differs from physicalism in that it takes phenomenal consciousness to be the only *irreducible* aspect of nature, as opposed to an epiphenomenon or emergent property of physical arrangements. It also differs from dualism in that it takes all physical elements and arrangements to exist *in consciousness*—solely as phenomenal properties—as opposed to outside consciousness.

Historically, idealism has had many different variations labeled as *subjective idealism, absolute idealism, actual idealism*, etc. It is not my purpose here to elaborate on the subtle, ambiguous and often contentious differences among these variations. Instead, I want to simply describe the basic tenets that any plausible, modern formulation of idealism must entail, *given our present knowledge and understanding* of the world. What follows is but a brief summary of a much more extensive derivation of idealism from first principles (Kastrup 2017b, 2017e[1]).

The defining tenet of idealism is the notion that all reality is in a *universal* form of consciousness—thus not bound to personal boundaries—arising as patterns of excitation *of* this universal consciousness. Our personal psyche forms through a process of dissociation in universal consciousness, analogous to how the psyche of a person suffering from dissociative identity disorder (DID) differentiates itself into multiple centers of experience called *alters* (Braude 1995, Kelly et al. 2009, Schlumpf et al. 2014). Recent research has demonstrated the literally *blinding* power of dissociation (Strasburger & Waldvogel 2015). This way, there is a sense in which each living creature is an alter of universal consciousness, which explains why we aren't aware of each

---

1 These references can be found in Chapters 5 and 6 of this volume, respectively.

other's inner lives or of what happens across time and space at a universal scale.

The formation of an alter in universal consciousness creates a boundary—a "Markov Blanket" (Friston, Sengupta & Auletta 2014: 430-432)—between phenomenality internal to the alter and that external to it. Phenomenality external to the alter—but still in its vicinity—impinges on the alter's boundary. The plausibility of this kind of phenomenal impingement from across a dissociative boundary is well established: we know, for instance, that dissociated feelings can dramatically affect our thoughts and, thereby, behaviors (Lynch & Kilmartin 2013), while dissociated expectations routinely mold our perceptions (cf. Eagleman 2011).

The impingement of external phenomenality on an alter's boundary is what we call sense perception. The world we perceive around ourselves is thus a *coded phenomenal representation* (Friston, Sengupta & Auletta 2014: 432-434)—which I shall call the *extrinsic appearance*—of equally phenomenal processes unfolding across the dissociative boundary of our alter.

A living biological body is the extrinsic appearance of an alter in universal consciousness. In particular, our sense organs—including our skin—are the extrinsic appearance of our alter's boundary. As such, our brain and its electrochemical activity are part of what our inner life *looks like* from across its dissociative boundary. Of course, both the extrinsic appearance and the corresponding inner life are phenomenal in nature. They are both experiences.

A person's brain activity correlates with the person's reported inner life because the former is but a coded representation of the latter. We all inhabit the same world because our respective alters are surrounded by the same universal field of phenomenality, like whirlpools in a single stream. And we can't change the patterns and regularities that govern the world—that is, the laws of nature—because our volition, as part of our alter, is dissociated from the rest of nature.

See Figure A.1 for a graphical depiction of all this.

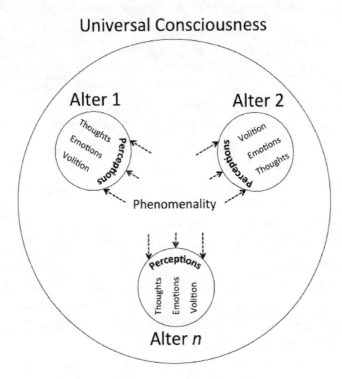

Figure A.1: Idealism in a nutshell.

Clearly, all three basic observations about reality discussed in Section A.2 can be rather simply explained by this parsimonious idealist ontology. Moreover, unlike physicalism and dualism, the ontology can also be reconciled with the empirical results discussed in Section A.3. It thus offers a more promising alternative for interpreting the relationship between body and consciousness than physicalism and dualism. The question that remains to be addressed is this: If idealism is true, what can we then infer about consciousness after bodily death? This is what the next section will attempt to answer.

A.5 What idealism says about consciousness after death

The idealist ontology briefly summarized in the previous section asserts that the physical body is the extrinsic appearance—the *image*—of a dissociative process in universal consciousness. In other words, a living body is what dissociation—meant simply descriptively, not as something negative or pathological—in universal consciousness *looks like*. Therefore, the death and ultimate dissolution of the body can only be the image of the *end of the dissociation*. Any other conclusion would violate the internal logic of idealism.

The reasoning here is rather straightforward but its implications profound. The hallmark of dissociation is "a disruption of and/or discontinuity in the normal integration of consciousness, memory, identity [and] emotion" (Black & Grant 2014: 191). Therefore, the end of dissociation can only entail a *reintegration* of "memory, identity [and] emotion" lost at birth. This means that bodily death, under idealism, must correlate with an *expansion* of our felt sense of identity, access to a broader set of memories and enrichment of our emotional inner life.

This conclusion is the exact opposite of what our mainstream physicalist ontology asserts. Moreover, there is nothing in the popular dualist alternative—mainly found in religious circles— that requires it either. So idealism is not only unique in its ability to explain reality more parsimoniously and completely than physicalism and dualism, it also offers a unique perspective on death.

Circumstantially but significantly, much of the literature regarding near-death experiences (NDEs) seems to corroborate this prediction of idealism (Kelly et al. 2009). To mention only one recent example, Anita Moorjani wrote of her felt sense of identity during her NDE: "I certainly don't feel reduced or smaller in any way. On the contrary, I haven't ever been this huge, this powerful, or this all-encompassing. ... [I] felt greater and more intense and expansive than my physical being" (2012: 69). It's hard to conceive of a more unambiguous confirmation of idealism's prediction than this passage, although Moorjani's entire NDE report echoes the prediction precisely.

Moreover, as recent studies have shown (Carhart-Harris et al. 2012, Palhano-Fontes et al. 2015, Carhart-Harris et al. 2016), psychedelic drugs *reduce* brain activity.[2] This suggests that psychedelic trances may be in some way akin to the early stages of the death process, offering glimpses into how death is experienced from a first-person perspective. And as we know, psychedelic trances do entail an unambiguous expansion of awareness (Strassman 2001, Griffiths et al. 2006, Strassman et al. 2008), which again seems to circumstantially corroborate idealism's prediction.[3]

A.6 Conclusions

To make educated guesses about what happens to consciousness upon bodily death, one has to have some understanding of the relationship between body and consciousness during life. This relationship, of course, reflects an ontology. So the question of

---

2 A later study performed at the University of Zürich has confirmed this further, showing that a psychedelic causes "significantly reduced absolute perfusion" (that is, blood flow) in just about every region of the brain, whilst leading to "profound subjective drug effects" (Lewis et al. 2017).

3 Although I only mention NDEs and psychedelic trances as evidence favoring idealism in this brief essay, all the phenomena discussed in Chapter 11 constitute further evidence, for exactly the same reason: they, too, show that impairment of brain function can sometimes correlate with expansion of awareness, just as predicted by idealism. There is nothing in physicalism or dualism that would render any of the evidence discussed in Chapter 11 expectable under these two ontologies.

what happens after death can be transposed into the question of which ontology is most plausible for making sense of the world during life.

While physicalism is our culture's academically-endorsed, mainstream ontology and dualism a popular alternative in religious circles, neither ontology seems tenable in view of recent experimental results in physics. Moreover, both ontologies suffer from problems such as lack of parsimony and limited explanatory power.

A third ontology, known as idealism, overcomes not only these problems but can also be reconciled with the available empirical evidence. It elegantly explains the three basic facts of reality: (a) that brain activity correlates with experience, (b) that we all seem to share the same world, and (c) that we can't change the laws of nature at will.

If idealism is correct, it implies that, instead of disappearing, conscious inner life *expands* — whatever new phenomenality this expansion may entail — upon bodily death. This prediction finds circumstantial but significant confirmation in reports of near-death experiences and psychedelic trances,[4] both of which can be construed as glimpses into the early stages of the death process.

---

4 See previous note.

Bibliography

Ackroyd, E. (1993). *A Dictionary of Dream Symbols*. London, UK: Cassell Illustrated.

Adelson, EH (1995). Checker shadow illusion. [Online]. Available from: http://persci.mit.edu/gallery/checkershadow [Accessed 12 April 2017].

Albert, H. (1985). *Treatise on Critical Reason*. Princeton, NJ: Princeton University Press.

American Psychiatric Association (2013). *Diagnostic and Statistical Manual of Mental Disorders* (5th ed.). Washington, DC: American Psychiatric Publishing.

Ananthaswamy, A. (2011). Quantum magic trick shows reality is what you make it. *New Scientist*, June 22. [Online]. Available from: https://www.newscientist.com/article/dn20600-quantum-magic-trick-shows-reality-is-what-you-make-it/ [Accessed 14 June 2016].

Aspect, A., Grangier, P. and Roger, G. (1981). Experimental Tests of Realistic Local Theories via Bell's Theorem. *Physical Review Letters*, 47 (7): 460-463.

Aspect, A., Dalibard, J. and Roger, G. (1982). Experimental Test of Bell's Inequalities Using Time-Varying Analyzers. *Physical Review Letters*, 49 (25): 1804-1807.

Aspect, A., Grangier, P. and Roger, G. (1982). Experimental Realization of Einstein-Podolsky-Rosen-Bohm Gedankenexperiment: A New Violation of Bell's Inequalities. *Physical Review Letters*, 49 (2): 91-94.

Augusto, LM (2010). Unconscious knowledge: A survey. *Advances in Cognitive Psychology*, 6: 116-141.

Baldwin, M. (2014). Is the Peer Review Process for Scientific Papers Broken? *Time*, April 29. [Online]. Available from: http://time.com/81388/is-the-peer-review-process-for-scientific-papers-broken/ [Accessed 6 August 2016].

Barfield, O. (2011). *Saving the Appearances: A Study in Idolatry.* Oxford, UK: Barfield Press.

Beall, J. (n.d.). Beall's list: Potential, possible, or probable predatory scholarly open-access publishers. *Scholarly Open Access*. [Online]. Originally available from: https://scholarlyoa.com/publishers/. An archived version can be found at: https://web.archive.org/web/20170112125427/https://scholarlyoa.com/publishers/ [Accessed 20 February 2017].

Bell, J. (1964). On the Einstein Podolsky Rosen Paradox. *Physics,* 1 (3): 195-200.

Black, DW and Grant, JE (2014). *The Essential Companion to the Diagnostic and Statistical Manual of Mental Disorders* (5th ed.). Washington, DC: American Psychiatric Publishing.

Blanke, O. et al. (2002). Stimulating illusory own-body perceptions: The part of the brain that can induce out-of-body experiences has been located. *Nature*, 419: 269-270.

Bohannon, J. (2013). Who's Afraid of Peer Review? *Science*, 342 (6154): 60-65.

Bohm, D. (1952a). A Suggested Interpretation of the Quantum Theory in Terms of 'Hidden' Variables. I. *Physical Review*, 85 (2): 166-179.

Bohm, D. (1952b). A Suggested Interpretation of the Quantum Theory in Terms of 'Hidden' Variables. II. *Physical Review*, 85 (2): 180-193.

Boly, M. et al. (2011). Preserved Feedforward But Impaired Top-Down Processes in the Vegetative State. *Science*, 332 (6031): 858-862.

Boswell, J. (1820). *The Life of Samuel Johnson, LL. D.* (Volume 1). London, UK: J. Davis, Military Chronicle and Military Classics Office.

Braude, S. (1995). *First Person Plural: Multiple Personality and the Philosophy of Mind*. New York, NY: Routledge.

Brenner, ED et al. (2006). Plant neurobiology: An integrated view of plant signaling. *Trends in Plant Science*, 11 (8): 413-419.

Burke, BL, Martens, A. and Faucher, EH (2010). Two Decades of Terror Management Theory: A Meta-Analysis of Mortality Salience Research. *Personality and Social Psychology Review*, 14 (2): 155-195.

Carhart-Harris, RL et al. (2012). Neural correlates of the psychedelic state as determined by fMRI studies with psilocybin. *Proceedings of the National Academy of Sciences of the United States of America*, 109 (6): 2138-2143.

Carhart-Harris, RL et al. (2016). Neural correlates of the LSD experience revealed by multimodal neuroimaging. *Proceedings of the National Academy of Sciences of the United States of America* (PNAS Early Edition), doi: 10.1073/pnas.1518377113.

Cartwright, J. (2007). Quantum physics says goodbye to reality. *IOP Physics World*, April 20. [Online]. Available from: http://physicsworld.com/cws/article/news/2007/apr/20/quantum-physics-says-goodbye-to-reality [Accessed 14 June 2016].

Catani, D. (2013). *Evil: A History in Modern French Literature and Thought*. London, UK: Bloomsbury.

Chalmers, D. (1996). *The Conscious Mind: In Search of a Fundamental Theory*. Oxford, UK: Oxford University Press.

Chalmers, D. (2003). Consciousness and its Place in Nature. In: Stich, S. & Warfield, T. (eds.). *Blackwell Guide to the Philosophy of Mind*. Malden, MA: Blackwell.

Chalmers, D. (2006). Strong and Weak Emergence. In: Davies, P. and Clayton, P. (eds.). *The Re-Emergence of Emergence*. Oxford, UK: Oxford University Press.

Chalmers, D. (2016). The Combination Problem for Panpsychism. In: Brüntrup, G. & Jaskolla, L. (eds.). *Panpsychism*. Oxford, UK: Oxford University Press.

Chalmers, D. (forthcoming). Idealism and the Mind-Body Problem. In: Seager, W. (ed.). *The Routledge Handbook of Panpsychism*. London, UK: Routledge.

Cheetham, T. (2012). *All the World an Icon: Henry Corbin and the*

Angelic Function of Beings. Berkeley, CA: North Atlantic Books.

Cleeremans, A. (2011). The radical plasticity thesis: How the brain learns to be conscious. *Frontiers in Psychology*, 2, article 86.

Coleman, S. (2014). The Real Combination Problem: Panpsychism, Micro-Subjects, and Emergence. *Erkenntnis*, 79 (1): 19-44.

Costandi, M. (2013). Brain scans decode dream content. *The Guardian*, April 5. [Online]. Available from: https://www.theguardian.com/science/neurophilosophy/2013/apr/05/brain-scans-decode-dream-content [Accessed 9 August 2016].

Cristofori, I. et al. (2016). Neural correlates of mystical experience. *Neuropsychologia*, 80: 212-220.

Damasio, A. (2011). Neural basis of emotions. *Scholarpedia*, 6 (3): 1804.

Dehaene, S. and Changeux, J-P (2011). Experimental and theoretical approaches to conscious processing. *Neuron*, 70 (2): 200-227.

Dennett, D. (1991). *Consciousness Explained*. London, UK: Penguin Books.

Dennett, D. (2003). Explaining the 'Magic' of Consciousness. *Journal of Cultural and Evolutionary Psychology*, 1 (1): 7-19.

Dick, PK (2001). *Valis*. London, UK: Gollancz.

Dijksterhuis, A. and Nordgren, LF (2006). A theory of unconscious thought. *Perspectives on Psychological Science*, 1 (2): 95-109.

DiSalvo, D. (2012). When You Inject Spirit Mediums' Brains with Radioactive Chemicals, Strange Things Happen. *Forbes*, November 18. [Online]. Available from: http://www.forbes.com/sites/daviddisalvo/2012/11/18/when-you-inject-spirit-mediums-brains-with-radioactive-chemicals-some-really-strange-things-happen/ [Accessed 25 February 2017].

Eagleman, D. (2011). *Incognito: The Secret Lives of the Brain*. New York, NY: Canongate.

Einstein, A., Podolsky, B. and Rosen, N. (1935). Can Quantum-Mechanical Description of Physical Reality be Considered

Complete? *Physical Review*, 47: 777-780.

Eliade, M. (2009). *Rites and Symbols of Initiation: The Mysteries of Birth and Rebirth*. New York, NY: Spring Publications.

Feynman, R. (1999). *The Pleasure of Finding Things Out*. Cambridge, MA: Perseus Publishing.

Fields, C. et al. (2017). Eigenforms, interfaces and holographic encoding: Toward an evolutionary account of objects and spacetime. *Constructivist Foundations*, 12 (3): 265-291.

Floridi, L. (2008). Trends in the philosophy of information. In: Adriaans, P. and Benthem, J. van (eds.). *Handbook of the Philosophy of Science, Volume 8: Philosophy of Information*. Amsterdam, The Netherlands: Elsevier, pp. 113-131.

Fonagy, P. et al. (eds.) (2012). *The Significance of Dreams*. London, UK: Karnac Books.

Ford, BJ (2010). The secret power of the single cell. *New Scientist*, 206 (2757): 26-27.

Fraassen, BC van (1980). *The Scientific Image*. Oxford, UK: Oxford University Press.

Fraassen, BC van (1990). *Laws and Symmetry*. Oxford, UK: Oxford University Press.

Frankl, VE (1991). *The Will to Meaning* (Expanded ed.). New York, NY: Meridian.

Franklin, SP (1997). *Artificial Minds*. Cambridge, MA: MIT Press.

Franz, M-L von (1964). The process of individuation. In: Jung, CG (ed.). *Man and His Symbols*. New York, NY: Anchor Press.

Franz, M-L von (1974). *Number and Time: Reflections Leading Toward a Unification of Depth Psychology and Physics*. Evanston, IL: Northwestern University Press.

Franz, M-L von and Boa, F. (1994). *The Way of the Dream*. Boston, MA: Shambhala Publications.

Fredkin, E. (2003). An Introduction to Digital Philosophy. *International Journal of Theoretical Physics*, 42 (2): 189-247.

Fredkin, E. (n.a.). *On the soul (draft)*. [Online]. http://www. digitalphilosophy.org//wp-content/uploads/2015/07/on_the_ soul.pdf [Accessed 2 July 2016].

Friston, K. (2013). Life as we know it. *Journal of the Royal Society Interface*, 10 (86): 20130475.

Friston, K., Sengupta, B. and Auletta, G. (2014). Cognitive Dynamics: From Attractors to Active Inference. *Proceedings of the IEEE*, 102 (4): 427-445.

Gaal, S. van et al. (2011). Dissociable Brain Mechanisms Underlying the Conscious and Unconscious Control of Behavior. *Journal of Cognitive Neuroscience*, 23 (1): 91-105.

Gabrielsen, P. (2013). When Does Your Baby Become Conscious? *Science*, April 18. [Online]. Available from: http://www.sciencemag.org/news/2013/04/when-does-your-baby-become-conscious [Accessed 10 September 2017].

Gers, F., Garis, H. and Korkin, M. (2005). CoDi-1Bit: A simplified cellular automata based neuron model. *Lecture Notes in Computer Science*, 1363: 315-333.

Gillespie, A. (2007). The Social Basis of Self-Reflection. In: Valsiner, J. and Rosa, A. (eds.). *The Cambridge Handbook of Sociocultural Psychology*. New York, NY: Cambridge University Press, pp. 678-691.

Glasersfeld, E. von (1987). An Introduction to Radical Constructivism. In: Watzlawick, P. (ed.). *The Invented Reality*. New York, NY: W. W. Norton & Company.

Godfrey-Smith, P. (2014). *Philosophy of Biology*. Princeton, NJ: Princeton University Press.

Goff, P. (2009). Why Panpsychism doesn't Help Us Explain Consciousness. *Dialectica*, 63 (3): 289-311.

Grangier, P. (2001). Contextual objectivity: A realistic interpretation of quantum mechanics. *arXiv:quant-ph/0 012122v2*. [Online]. Available from: https://arxiv.org/abs/quant-ph/0012122 [Accessed 4 September 2017].

Graziano, M. (2016). Consciousness Is Not Mysterious. *The Atlantic*, January 12. [Online]. Available from: http://www.theatlantic.com/science/archive/2016/01/consciousness-color-brain/423522/ [Accessed 26 February 2017].

Greene, B. (2003). *The Elegant Universe: Superstrings, Hidden Dimensions, and the Quest for the Ultimate Theory*. New York, NY: W.W. Norton & Company.

Griffin, D. (1998). *Unsnarling the World-Knot*. Eugene, OR: Wipf & Stock.

Griffiths, RR et al. (2006). Psilocybin can occasion mystical-type experiences having substantial and sustained personal meaning and spiritual significance. *Psychopharmacology*, 187 (3): 268-283.

Gröblacher, S. et al. (2007). An experimental test of non-local realism. *Nature*, 446: 871-875.

Haikonen, P. (2003). *The Cognitive Approach to Conscious Machines*. Exeter, UK: Imprint Academic.

Haikonen, P. (2007). *Robot Brains: Circuits and Systems for Conscious Machines*. Chichester, UK: John Wiley & Sons.

Hameroff, S. (2006). Consciousness, Neurobiology and Quantum Mechanics: The Case for a Connection. In: Tuszynski, J. (ed.). *The Emerging Physics of Consciousness*. Berlin, Germany: Springer Verlag.

Harris, S. (2012a). *Free Will*. New York, NY: Free Press.

Harris, S. (2012b). Science on the Brink of Death. *samharris.org*, 11 November. [Online]. Available from: http://www.samharris.org/blog/item/science-on-the-brink-of-death [Accessed 25 March 2017].

Harris, S. (2016). You are more than your brain! *Big Think*, September 4. [Online]. https://www.facebook.com/BigThinkdotcom/videos/10153879575418527/ [Accessed 1 November 2016].

Hart, J. (2013). Toward an Integrative Theory of Psychological Defense. *Perspectives on Psychological Science*, 9 (1): 19-39.

Hart, DB (2017). The Illusionist. *The New Atlantis*, Summer/Fall issue: 109-121.

Hassin, R., Uleman, J. and Bargh, J. (eds.) (2005). *The New*

Unconscious. New York, NY: Oxford University Press.

Hassin, RR (2013). Yes It Can: On the Functional Abilities of the Human Unconscious. *Perspectives on Psychological Science*, 8 (2): 195-207.

Heflick, NA et al. (2015). Death awareness and body-self dualism: A why and how of afterlife belief. *European Journal of Social Psychology*, 45 (2): 267-275.

Heine, SJ, Proulx, T. and Vohs, KD (2006). The Meaning Maintenance Model: On the Coherence of Social Motivations. *Personality and Social Psychology Review*, 10 (2): 88-110.

Henry, RC (2005). The mental Universe. *Nature*, 436: 29.

Hensen, B. et al. (2015). Loophole-free Bell inequality violation using electron spins separated by 1.3 kilometres. *Nature*, 526: 682-686.

Hilgard, E. (1977). *Divided Consciousness*. New York, NY: John Wiley & Sons.

Hoffman, DD (2009). The Interface Theory of Perception: Natural Selection Drives True Perception to Swift Extinction. In: Dickinson, S. et al. (eds.). *Object Categorization: Computer and Human Vision Perspectives*. Cambridge, UK: Cambridge University Press.

Hoffman, DD and Singh, M. (2012). Computational Evolutionary Perception. *Perception*, 41 (9): 1073-1091.

Horgan, T. and Potrč, M. (2000). Blobjectivism and Indirect Correspondence. *Facta Philosophica*, 2 (2): 249-270.

Horikawa, T. et al. (2013). Neural Decoding of Visual Imagery During Sleep. *Science*, 4, doi: 10.1126/science.1234330.

Husserl, E. (1970). *The Crisis of European Sciences and Transcendental Phenomenology: An Introduction to Phenomenological Philosophy*. Evanston, IL: Northwestern University Press.

Huyghe, P. and Parreno, P. (2003). *No Ghost Just a Shell*. Cologne, Germany: Verlag der Buchhandlung Walther König.

Immordino-Yang, MH et al. (2009). Neural Correlates of Admiration and Compassion. *Proceedings of the National*

Academy of Sciences of the United States of America, 106 (19): 8021-8026.

Jung, CG (1991). *The Archetypes and the Collective Unconscious* (2nd ed.). London, UK: Routledge.

Jung, CG (1995). *Memories, Dreams, Reflections*. London, UK: Fontana Press.

Jung, CG (2001). *Modern Man in Search of a Soul*. New York, NY: Routledge.

Jung, CG (2002). *Dreams*. London, UK: Routledge.

Jung, CG (2014). *Analytical Psychology: Its Theory and Practice* (2nd ed.). London, UK: Routledge.

Kafatos, M. and Nadeau, R. (1990). *The Conscious Universe: Part and Whole in Modern Physical Theory*. New York, NY: Springer-Verlag.

Kang, YHR et al. (2017). Piercing of Consciousness as a Threshold-Crossing Operation. *Current Biology*, doi:10.1016/j.cub.2017.06.047. [Online]. Available from: http://www.cell.com/current-biology/fulltext/S0960-9822(17)30784-4 [Accessed 4 August 2017].

Karunamuni, ND (2015). The Five-Aggregate Model of the Mind. *SAGE Open*, 5 (2), doi:10.1177/2158244015583860.

Kastrup, B. (2014). *Why Materialism Is Baloney*. Winchester, UK: Iff Books.

Kastrup, B. (2015). *Brief Peeks Beyond*. Winchester, UK: Iff Books.

Kastrup, B. (2016a). *More Than Allegory*. Winchester, UK: Iff Books.

Kastrup, B. (2016b). What Neuroimaging of the Psychedelic State Tells Us about the Mind-Body Problem. *Journal of Cognition and Neuroethics*, 4 (2): 1-9.

Kastrup, B. (2017a). Self-Transcendence Correlates with Brain Function Impairment. *Journal of Cognition and Neuroethics*, 4 (3): 33-42.

Kastrup, B. (2017b). An Ontological Solution to the Mind-Body Problem. *Philosophies*, 2 (2), doi:10.3390/philosophies2020010.

Kastrup, B. (2017c). On the Plausibility of Idealism: Refuting Criticisms. *Disputatio*, 9 (44): 13-34.

Kastrup, B. (2017d). There Is an 'Unconscious,' but It May Well Be Conscious. *Europe's Journal of Psychology*, 13 (3): 559-572.

Kastrup, B. (2017e). Making Sense of the Mental Universe. *Philosophy and Cosmology*, 19: 33-49.

Kay, AC et al. (2010). Religious belief as compensatory control. *Personality and Social Psychology Review*, 14 (1): 37-48.

Kelly, EF et al. (2009). *Irreducible Mind: Toward a Psychology for the 21st Century*. Lanham, MD: Rowman & Littlefield.

Kihlstrom, J. and Cork, R. (2007). Anesthesia. In: Velmans, M. & Schneider, S. (eds.). *The Blackwell Companion to Consciousness*. Oxford, UK: Blackwell.

Kim, Y-H et al. (2000). A Delayed 'Choice' Quantum Eraser. *Physical Review Letters*, 84: 1-5.

Klein, SB (2015). The Feeling of Personal Ownership of One's Mental States: A Conceptual Argument and Empirical Evidence for an Essential, but Underappreciated, Mechanism of Mind. *Psychology of Consciousness: Theory, Research, and Practice*, 2 (4): 355-376.

Klimov, PV et al. (2015). Quantum entanglement at ambient conditions in a macroscopic solid-state spin ensemble. *Science Advances*, 1 (10), e1501015.

Koch, C. (2004). *The Quest for Consciousness: A Neurobiological Approach*. Englewood, CO: Roberts & Company Publishers.

Koch, C. (2012a). *Consciousness: Confessions of a Romantic Reductionist*. Cambridge, MA: MIT Press.

Koch, C. (2012b). This is Your Brain on Drugs. *Scientific American Mind*, 1 May. [Online]. Available from: http://www.scientificamerican.com/article/this-is-your-brain-on-drugs/ [Accessed 9 August 2016].

Krioukov, D. et al. (2012). Network Cosmology. *Scientific Reports*, 2: doi:10.1038/srep00793.

Kuhn, T. (1996). *The Structure of Scientific Revolutions* (3rd ed.).

Chicago, IL: University of Chicago Press.

Kurzweil, R. (2005). *The Singularity Is Near*. New York, NY: Viking.

Landau, MJ et al. (2004). A Function of Form: Terror Management and Structuring the Social World. *Journal of Personality and Social Psychology*, 87 (2): 190-210.

Langer, E. and Rodin, J. (1976). The effects of choice and enhanced personal responsibility for the aged: a field experiment in an institutional setting. *Journal of Personality and Social Psychology*, 34 (2): 191-198.

Lapkiewicz, R. et al. (2011). Experimental non-classicality of an indivisible quantum system. *Nature*, 474: 490-493.

Lee, KC et al. (2011). Entangling Macroscopic Diamonds at Room Temperature. *Science*, 334 (6060): 1253-1256.

Leggett, AN (2003). Nonlocal Hidden-Variable Theories and Quantum Mechanics: An Incompatibility Theorem. *Foundations of Physics*, 33 (10): 1469-1493.

Levine, J. (1983). Materialism and qualia: The explanatory gap. *Pacific Philosophical Quarterly*, 64: 354-361.

Lewis, CR et al. (2017). Two dose investigation of the 5-HT-agonist psilocybin on relative and global cerebral blood flow. *NeuroImage*, July, doi:10.1016/j.neuroimage.2017.07.020.

Libet, B. (1985). Unconscious cerebral initiative and the role of conscious will in voluntary action. *Behavioral and Brain Sciences*, 8: 529-566.

Linde, A. (1998). *Universe, Life, Consciousness*. A paper delivered at the Physics and Cosmology Group of the "Science and Spiritual Quest" program of the Center for Theology and the Natural Sciences (CTNS), Berkeley, California. [Online]. Available from: web.stanford.edu/~alinde/SpirQuest.doc [Accessed 14 June 2016].

Lommel, P. van et al. (2001). Near-death experience in survivors of cardiac arrest: a prospective study in the Netherlands. *The Lancet*, 358 (9298): 2039-2045.

Luck, A. et al. (1999). Effects of video information on precolonoscopy anxiety and knowledge: a randomised trial. *The Lancet*, 354 (9195): 2032-2035.

Lynch, J. and Kilmartin, C. (2013). *Overcoming Masculine Depression: The Pain Behind the Mask*. New York, NY: Routledge.

Lythgoe, M. et al. (2005). Obsessive, prolific artistic output following subarachnoid hemorrhage. *Neurology*, 64 (2): 397-398.

Ma, X-S et al. (2013). Quantum erasure with causally disconnected choice. *Proceedings of the National Academy of Sciences of the Unites States of America*, 110 (4): 1221-1226.

Macnab, AJ et al. (2009). Asphyxial games or "the choking game": A potentially fatal risk behaviour. *Injury Prevention*, 15 (1): 45-49.

Maharaj, N. (1973). *I Am That*. Mumbai, India: Chetana Publishing.

Manning, AG et al. (2015). Wheeler's delayed-choice gedanken experiment with a single atom. *Nature Physics*, 11: 539-542.

McCook, A. (2006). Is Peer Review Broken? *The Scientist*, February. [Online]. Available from: http://www.the-scientist.com/?articles.view/articleNo/23672/title/Is-Peer-Review-Broken-/ [Accessed 25 February 2017].

Mead, GRS (translator) (2010). *The Corpus Hermeticum*. Whitefish, MT: Kessinger Publishing LLC.

Merali, Z. (2015). Quantum 'spookiness' passes toughest test yet. *Nature*, News, August 27. [Online]. Available from: http://www.nature.com/news/quantum-spookiness-passes-toughest-test-yet-1.18255 [Accessed 30 August 2015].

Merleau-Ponty, M. (1964). *The Primacy of Perception: And Other Essays on Phenomenological Psychology, the Philosophy of Art, History and Politics*. Evanston, IL: Northwestern University Press.

Miller, B. et al. (1998). Emergence of artistic talent in frontotemporal dementia. *Neurology*, 51 (4): 978-982.

Miller, B. et al. (2000). Functional correlates of musical and visual ability in frontotemporal dementia. *The British Journal of Psychiatry*, 176: 458-463.

Moore, T. (2012). *Care of the Soul: An Inspirational Programme to Add Depth and Meaning to Your Everyday Life*. London, UK: Piatkus Books.

Moorjani, A. (2012). *Dying To Be Me: My Journey from Cancer, to Near Death, to True Healing*. Carlsbad, CA: Hay House.

Nagasawa, Y. and Wager, K. (2016). Panpsychism and Priority Cosmopsychism. In: Brüntrup, G. and Jaskolla, L. (eds.). *Panpsychism*. Oxford, UK: Oxford University Press.

Nagel, T. (1974). What is it like to be a bat? *The Philosophical Review*, 83 (4): 435-450.

Nagel, T. (2012). *Mind and Cosmos: Why the Materialist Neo-Darwinian Conception of Nature Is Almost Certainly False*. Oxford, UK: Oxford University Press.

Neal, RM (2008). *The Path to Addiction: "And Other Troubles We Are Born to Know."* Bloomington, IN: AuthorHouse.

Neumann, J. von (1996). *Mathematical Foundations of Quantum Mechanics*. Princeton, NJ: Princeton University Press.

Newell, BR and Shanks, DR (2014). Unconscious influences on decision making: A critical review. *Behavioral and Brain Sciences*, 37 (1): 1-19.

Nixon, GM (2010). From Panexperientialism to Conscious Experience: The Continuum of Experience. *Journal of Consciousness Exploration and Research*, 1 (3): 215-233.

Okasha, S. (2002). *Philosophy of Science: A Very Short Introduction*. Oxford, UK: Oxford University Press.

Ortiz-Osés, A. (2008). *The Sense of the World*. Aurora, CO: The Davies Group Publishers.

Palhano-Fontes, F. et al. (2015). The Psychedelic State Induced by Ayahuasca Modulates the Activity and Connectivity of the Default Mode Network. *PLoS ONE*, 10 (2): e0118143.

Paller, KA and Suzuki, S. (2014). The source of consciousness.

Trends in Cognitive Sciences, 18 (8): 387-389.

Partington, CF (ed.) (1837). *The British Cyclopædia of Natural History: A Scientific Classification of Animals, Plants, and Minerals; With a Popular View of Their Habits, Economy, and Structure*, Vol. 3. London, UK: WS Orr & Co., Amen Corner, Paternoster-Row.

Pearl, J. (1988). *Probabilistic Reasoning in Intelligent Systems: Networks of Plausible Inference*. San Francisco, CA: Morgan Kaufmann.

Peres, J. et al. (2012). Neuroimaging during Trance State: A Contribution to the Study of Dissociation. *PLoS ONE*, 7 (11): e49360.

Piccinini, G. (2015). Computation in physical systems. In: Zalta, EN (ed.). *The Stanford Encyclopedia of Philosophy* (Summer 2015 Edition). [Online]. Available from: http://plato.stanford.edu/archives/sum2015/entries/computation-physicalsystems [Accessed 30 June 2016].

Piore, A. (2013). The genius within. *Popular Science*, March: 46-53.

Pyszczynski, T., Greenberg, J. and Solomon, S. (1997). Why Do We Need What We Need? A Terror Management Perspective on the Roots of Human Social Motivation. *Psychological Inquiry*, 8 (1): 1-20.

Reddy, S. and Knight, K. (2011). What We Know About The Voynich Manuscript. In: *Proceedings of the 5th ACL-HLT Workshop on Language Technology for Cultural Heritage, Social Sciences, and Humanities*. Stroudsburg, PA: Association for Computational Linguistics, pp. 78-86.

Retz (2007). Tripping without drugs: Experience with hyperventilation (ID 14651). *Erowid.org*. [Online]. Available from: http://www.erowid.org/exp/14651 [Accessed 25 February 2017].

Rhinewine, JP and Williams, OJ (2007). Holotropic Breathwork: The Potential Role of a Prolonged, Voluntary Hyperventilation

Procedure as an Adjunct to Psychotherapy. *The Journal of Alternative and Complementary Medicine,* 13 (7): 771-776.

Robinson, H. (2016). Dualism. In: Zalta, EN (ed.). *The Stanford Encyclopedia of Philosophy* (Spring 2016 Edition). [Online]. Available from: http://plato.stanford.edu/archives/spr2016/entries/dualism [Accessed 17 June 2016].

Romero, J. et al. (2010). Violation of Leggett inequalities in orbital angular momentum subspaces. *New Journal of Physics,* 12: 123007. [Online]. Available from: http://iopscience.iop.org/article/10.1088/1367-2630/12/12/123007 [Accessed 14 June 2016].

Ronnenberg, A. and Martin, K. (2010). *The Book of Symbols.* Cologne, Germany: Taschen.

Rosenberg, G. (2004). *A Place for Consciousness.* New York, NY: Oxford University Press.

Rosner, RI, Lyddon, WJ and Freeman, A. (eds.) (2004). *Cognitive Therapy and Dreams.* New York, NY: Springer Publishing Company.

Ross, WD (1951). *Plato's Theory of Ideas.* Oxford, UK: Oxford University Press.

Rovelli, C. (2008). Relational Quantum Mechanics. *arXiv:quant-ph/9609002v2.* [Online]. Available from: https://arxiv.org/abs/quant-ph/9609002v2 [Accessed 1 August 2017].

Russell, B. (2009). *Human Knowledge: Its Scope and Limits.* London, UK: Routledge Classics.

Ryle, G. (2009). *The Concept of Mind.* London, UK: Routledge.

Sacks, O. (1985). *The Man Who Mistook His Wife for a Hat.* New York, NY: Harper & Row.

Sandberg, A. and Boström, N. (2008). *Whole Brain Emulation: A Roadmap* (Technical Report #2008-3). Oxford, UK: Future of Humanity Institute, Oxford University. [Online]. Available from: http://www.fhi.ox.ac.uk/brain-emulation-roadmap-report.pdf [Accessed 26 February 2017].

Sartre, J-P (1992). *Being and Nothingness: A Phenomenological Essay on Ontology*. New York, NY: Washington Square Press.

Schaffer, J. (2010). Monism: The Priority of the Whole. *Philosophical Review*, 119 (1): 31-76.

Schlumpf, Y. et al. (2014). Dissociative Part-Dependent Resting-State Activity in Dissociative Identity Disorder: A Controlled fMRI Perfusion Study. *PLoS ONE*, 9, doi:10.1371/journal. pone.0098795.

Schooler, J. (2002). Re-representing consciousness: dissociations between experience and meta-consciousness. *Trends in Cognitive Science*, 6 (8): 339-344.

Searle, JR (2004). *Mind: A Brief Introduction*. Oxford, UK: Oxford University Press.

Sellars, W. (1997). *Empiricism and the Philosophy of Mind*. Cambridge, MA: Harvard University Press.

Shani, I. (2015). Cosmopsychism: A Holistic Approach to the Metaphysics of Experience. *Philosophical Papers*, 44 (3): 389-437.

Shannon, CE (1948). A Mathematical Theory of Communication. *Bell System Technical Journal*, 27: 379-423 & 623-656.

Shermer, M. (2011). What Is Pseudoscience? *Scientific American*, September 1. [Online]. Available from: http://www.scientificamerican.com/article.cfm?id=what-is-pseudoscience [Accessed 7 August 2016].

Siegel, E. (2016). Ask Ethan: Is The Universe Itself Alive? *Forbes*, January 23. [Online]. Available from: http://www.forbes.com/sites/startswithabang/2016/01/23/ask-ethan-is-the-universe-itself-alive [Accessed 25 February 2017].

Skrbina, D. (2007). *Panpsychism in the West*. Cambridge, MA: MIT Press.

Smith, R. (2006). Peer review: a flawed process at the heart of science and journals. *Journal of the Royal Society of Medicine*, 99 (4): 178-182.

Smolin, L. (2007). *The Trouble With Physics: The Rise of String Theory, the Fall of a Science, and What Comes Next*. New York,

NY: Mariner Books.

Stannard, DE (1980). *Shrinking History: On Freud and the Failure of Psychohistory*. Oxford, UK: Oxford University Press.

Stavrova, O., Ehlebracht, D. and Fetchenhauer, D. (2016). Belief in scientific–technological progress and life satisfaction: The role of personal control. *Personality and Individual Differences*, 96: 227-236.

Stoerig, P. and Cowey, A. (1997). Blindsight in man and monkey. *Brain*, 120 (3): 535-559.

Stoljar, D. (2016). Physicalism. In: Zalta, EN (ed.). *The Stanford Encyclopedia of Philosophy* (Spring 2016 Edition). [Online]. Available from: http://plato.stanford.edu/archives/spr2016/entries/physicalism/ [Accessed 26 February 2017].

Strasburger, H. and Waldvogel, B. (2015). Sight and blindness in the same person: Gating in the visual system. *PsyCh Journal*, 4 (4): 178-185.

Strassman, R. (2001). *DMT: The Spirit Molecule*. Rochester, VT: Park Street Press.

Strassman, R. et al. (2008). *Inner Paths to Outer Space*. Rochester, VT: Park Street Press.

Strawson, G. et al. (2006). *Consciousness and Its Place in Nature*. Exeter, UK: Imprint Academic.

Swedenborg, E. (2007). *Heaven and its Wonders and Hell*. Charleston, SC: BiblioBazaar.

Tarlaci, S. and Pregnolato, M. (2016). Quantum neurophysics: From non-living matter to quantum neurobiology and psychopathology. *International Journal of Psychophysiology*, 103: 161-173.

Tarnas, R. (2010). *The Passion of the Western Mind*. London, UK: Pimlico.

Taylor, C. (2007). *A Secular Age*. Cambridge, MA: Harvard University Press.

Taylor, JB (2009). *My Stroke of Insight: A Brain Scientist's Personal Journey*. New York, NY: Viking.

Taylor, K. (1994). *The Breathwork Experience: Exploration and Healing in Nonordinary States of Consciousness*. Santa Cruz, CA: Hanford Mead.

Tegmark, M. (2014). *Our Mathematical Universe: My Quest for the Ultimate Nature of Reality*. New York, NY: Vintage Books.

Tillich, P. (1952). *The Courage To Be*. New Haven, CT: Yale University Press.

Tittel, W. et al. (1998). Violation of Bell Inequalities by Photons More Than 10 km Apart. *Physical Review Letters*, 81 (17): 3563-3566.

Tongeren, DR van and Green, JD (2010). Combating Meaninglessness: On the Automatic Defense of Meaning. *Personality and Social Psychology Bulletin*, 36 (10): 1372-1384.

Tononi, G. (2004). An information integration theory of consciousness. *BMC Neuroscience*, 5 (42): doi: 10.1186/1471-2202-5-42.

Treffert, D. (2006). *Extraordinary People: Understanding Savant Syndrome*. Omaha, NE: iUniverse, Inc.

Treffert, D. (2009). The savant syndrome: an extraordinary condition. A synopsis: past, present, future. *Philosophical Transactions of the Royal Society B*, 364 (1522): 1351-1357.

Tsuchiya, N. et al. (2015). No-Report Paradigms: Extracting the True Neural Correlates of Consciousness. *Trends in Cognitive Science*, 19 (12): 757-770.

University of California at Los Angeles (n.d.). *Lecture Demonstration Manual: Chladni Plate*. [Online]. Available from: http://demoweb.physics.ucla.edu/content/60-chladni-plate [Accessed 10 June 2017].

Urgesi, C. et al. (2010). The Spiritual Brain: Selective Cortical Lesions Modulate Human Self-Transcendence. *Neuron*, 65: 309-319.

Vaillant, GE (1992). *Ego Mechanisms of Defense: A Guide for Clinicians and Researchers*. Washington, DC: American Psychiatric Press.

Valsiner, J. (1998). *The Guided Mind*. Cambridge, MA: Harvard University Press.

Vandenbroucke, A. et al. (2014). Seeing without knowing: Neural Signatures of Perceptual Inference in the Absence of Report. *Journal of Cognitive Neuroscience*, 26 (5): 955-969.

Varela, FJ, Thompson, E. and Rosch, E. (1993). *The Embodied Mind: Cognitive Science and Human Experience*. Cambridge, MA: MIT Press.

Vazza, F. and Feletti, A. (2017). The Strange Similarity of Neuron and Galaxy Networks: Your life's memories could, in principle, be stored in the universe's structure. *Nautilus*, July 20. [Online]. Available from: http://nautil.us/issue/50/emergence/the-strange-similarity-of-neuron-and-galaxy-networks [Accessed 21 July 2017].

Walls, LD (2003). *Emerson's Life in Science: The Culture of Truth*. Ithaca, NY: Cornell University Press.

Watts, A. (1989). *The Book: On the Taboo Against Knowing Who You Are*. New York, NY: Vintage Books.

Webster, R. (1995). *Why Freud Was Wrong: Sin, Science, and Psychoanalysis*. New York, NY: Basic Books.

Wegner, DM (2002). *The Illusion of Conscious Will*. Cambridge, MA: MIT Press.

Weihs, G. et al. (1998). Violation of Bell's Inequality under Strict Einstein Locality Conditions. *Physical Review Letters*, 81 (23): 5039-5043.

Westen, D. (1999). The Scientific Status of Unconscious Processes: Is Freud Really Dead? *Journal of the American Psychoanalytic Association*, 47 (4): 1061-1106.

Whinnery, J. and Whinnery, A. (1990). Acceleration-Induced Loss of Consciousness: A Review of 500 Episodes. *Archives of Neurology*, 47 (7): 764-776.

Whitehead, AN (1947). *Essays in Science and Philosophy*. New York, NY: Philosophical Library.

Wigner, E. (1960). The Unreasonable Effectiveness of Mathematics in the Natural Sciences. *Communications on Pure and Applied Mathematics*, 13 (1): 1-14.

Windt, JM and Metzinger, T. (2007). The philosophy of dreaming and self-consciousness: what happens to the experiential subject during the dream state? In: Barrett, D. and McNamara, P. (eds.). *The New Science of Dreaming*. Westport, CT: Praeger, pp. 193-247.

Windt, J., Nielsen, T. and Thompson, E. (2016). Does Consciousness Disappear in Dreamless Sleep? *Trends in Cognitive Sciences*, 20 (12): 871-882.

Yetter-Chappell, H. (forthcoming). Idealism Without God. In: Goldschmidt, T. and Pearce, K. (eds.). *Idealism: New Essays in Metaphysics*. Oxford, UK: Oxford University Press.

Zemach, E. (2006). Wittgenstein's Philosophy of the Mystical. In: Copi, IM and Beard, RW (eds.). *Essays on Wittgenstein's Tractatus*. London, UK: Routledge, pp. 359-376.

Zicheng, H. (2006). *Vegetable Roots Discourse: Wisdom from Ming China on Life and Living: Caigentan*. Berkeley, CA: Shoemaker & Hoard.

Zuckerman Institute (2017). In Witnessing the Brain's 'Aha!' Moment, Scientists Shed Light on Biology of Consciousness. *Neuroscience News*, July 27. [Online]. Available from: http://neurosciencenews.com/consciousness-neuroscience-7189/ [Accessed 4 August 2017].

Zurek, WH (1994). Preferred Observables, Predictability, Classicality, and the Environment-Induced Decoherence. *arXiv:gr-qc/9402011v1*. [Online]. Available from: https://arxiv.org/abs/gr-qc/9402011 [Accessed 4 September 2017].

BOOKS

ACADEMIC AND SPECIALIST

Iff Books publishes non-fiction. It aims to work with authors and titles that augment our understanding of the human condition, society and civilisation, and the world or universe in which we live.
If you have enjoyed this book, why not tell other readers by posting a review on your preferred book site.
Recent bestsellers from Iff Books are:

Why Materialism Is Baloney
How True Skeptics Know There is no Death and Fathom Answers to Life, the Universe, and Everything
Bernardo Kastrup
A hard-nosed, logical, and skeptic non-materialist metaphysics, according to which the body is in mind, not mind in the body.
Paperback: 978-1-78279-362-5 ebook: 978-1-78279-361-8

The Fall
Steve Taylor
The Fall discusses human achievement versus the issues of war, patriarchy and social inequality.
Paperback: 978-1-90504-720-8 ebook: 978-184694-633-2